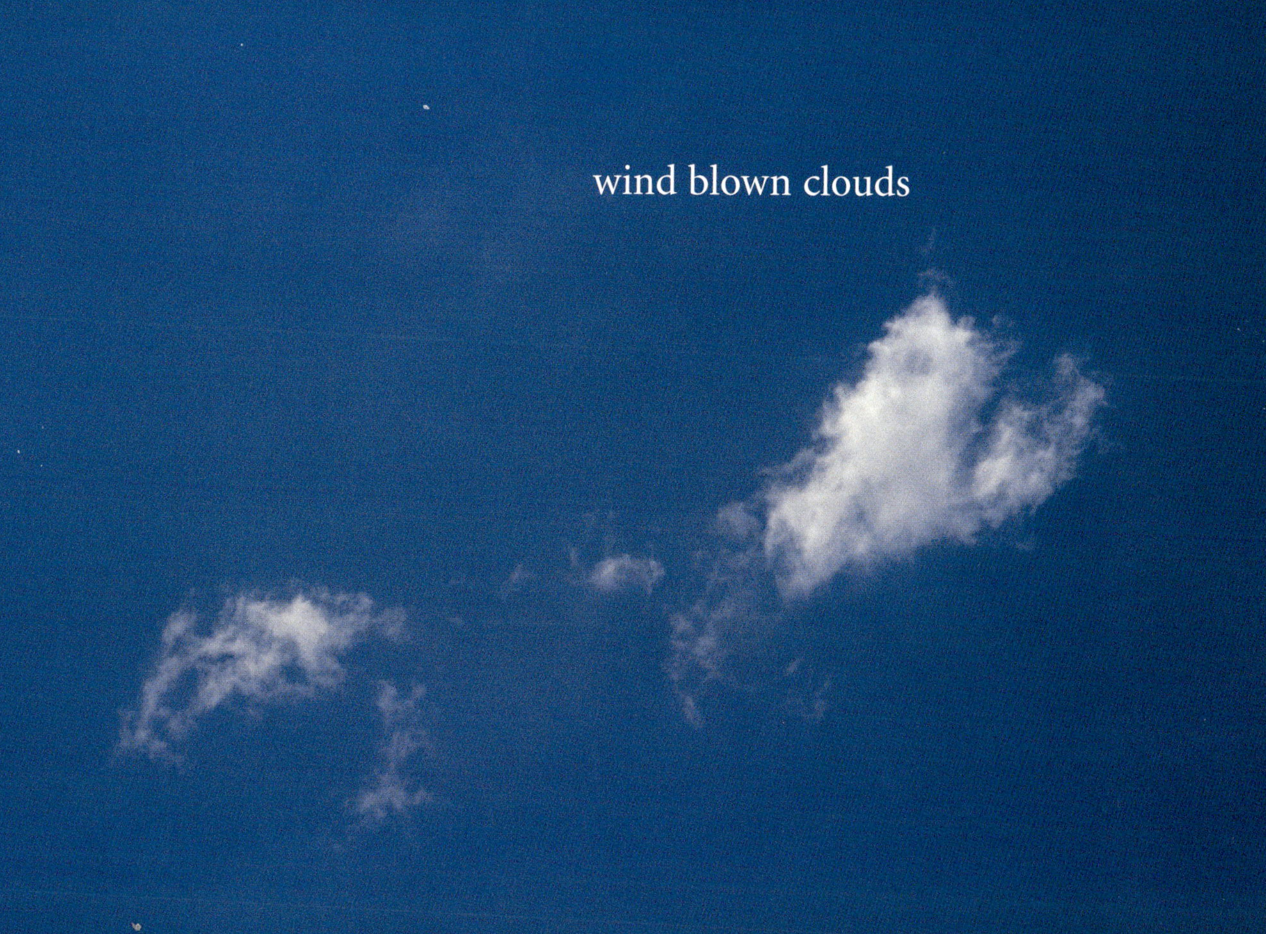
wind blown clouds

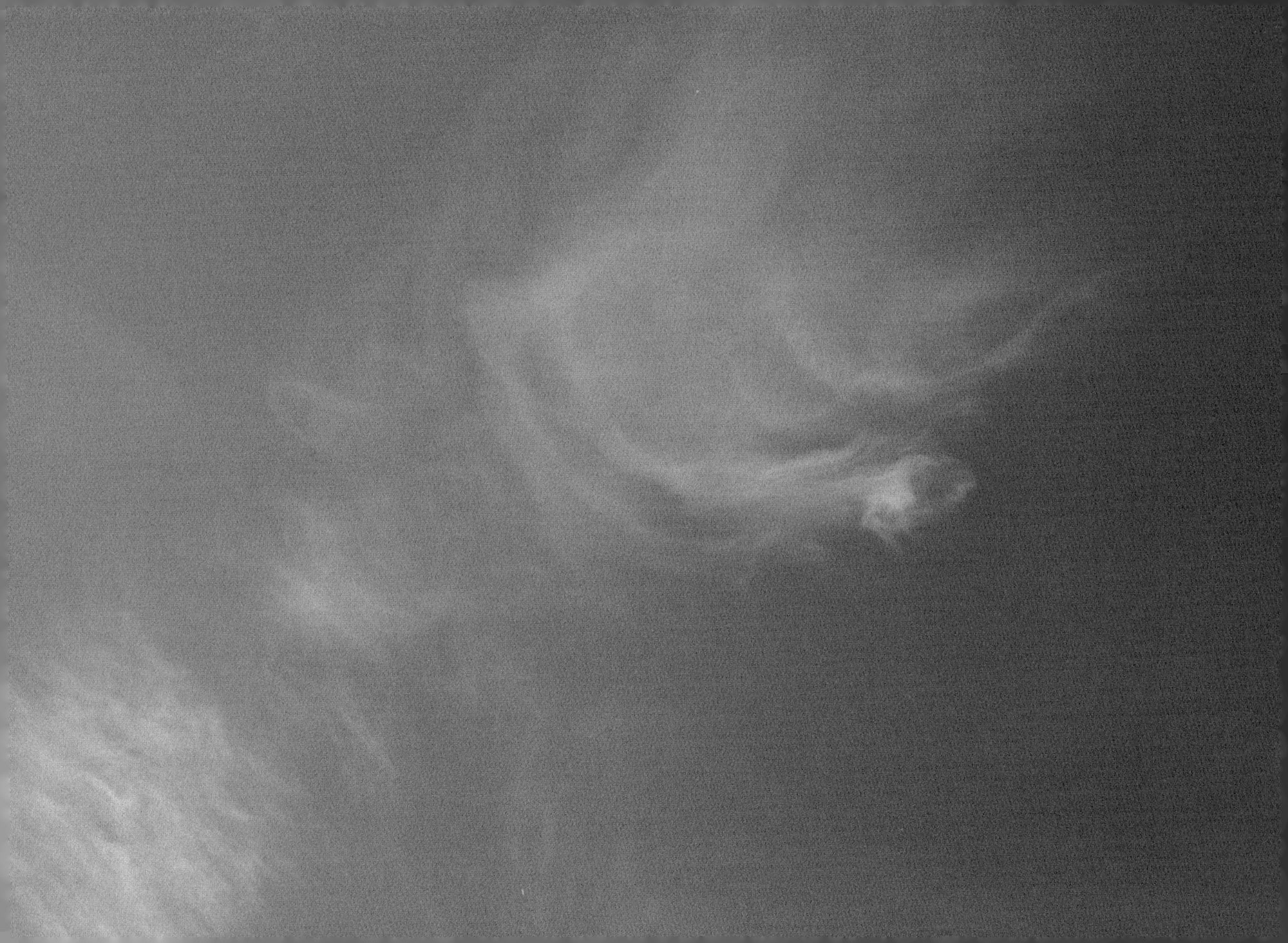

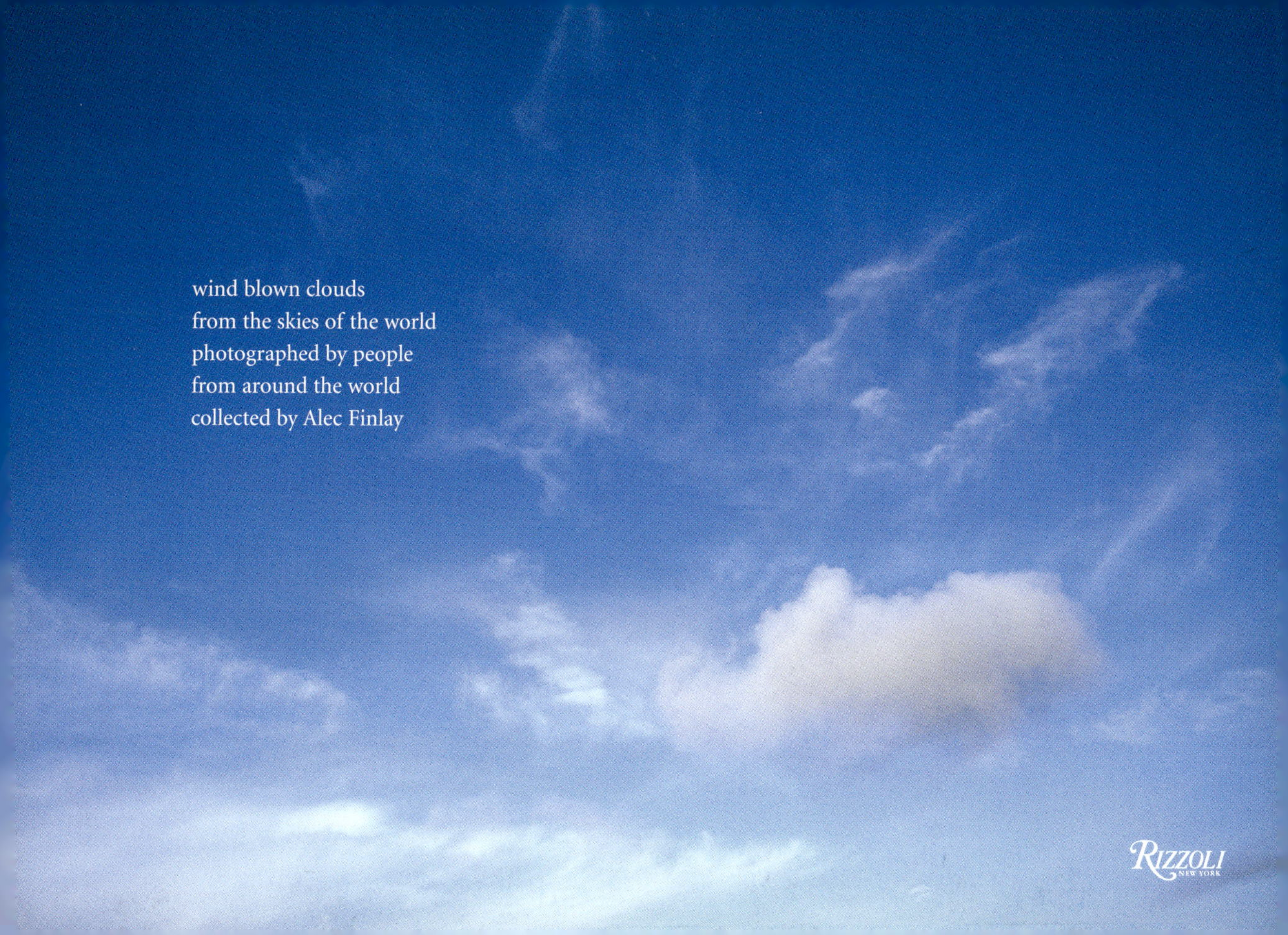

wind blown clouds
from the skies of the world
photographed by people
from around the world
collected by Alec Finlay

RIZZOLI
NEW YORK

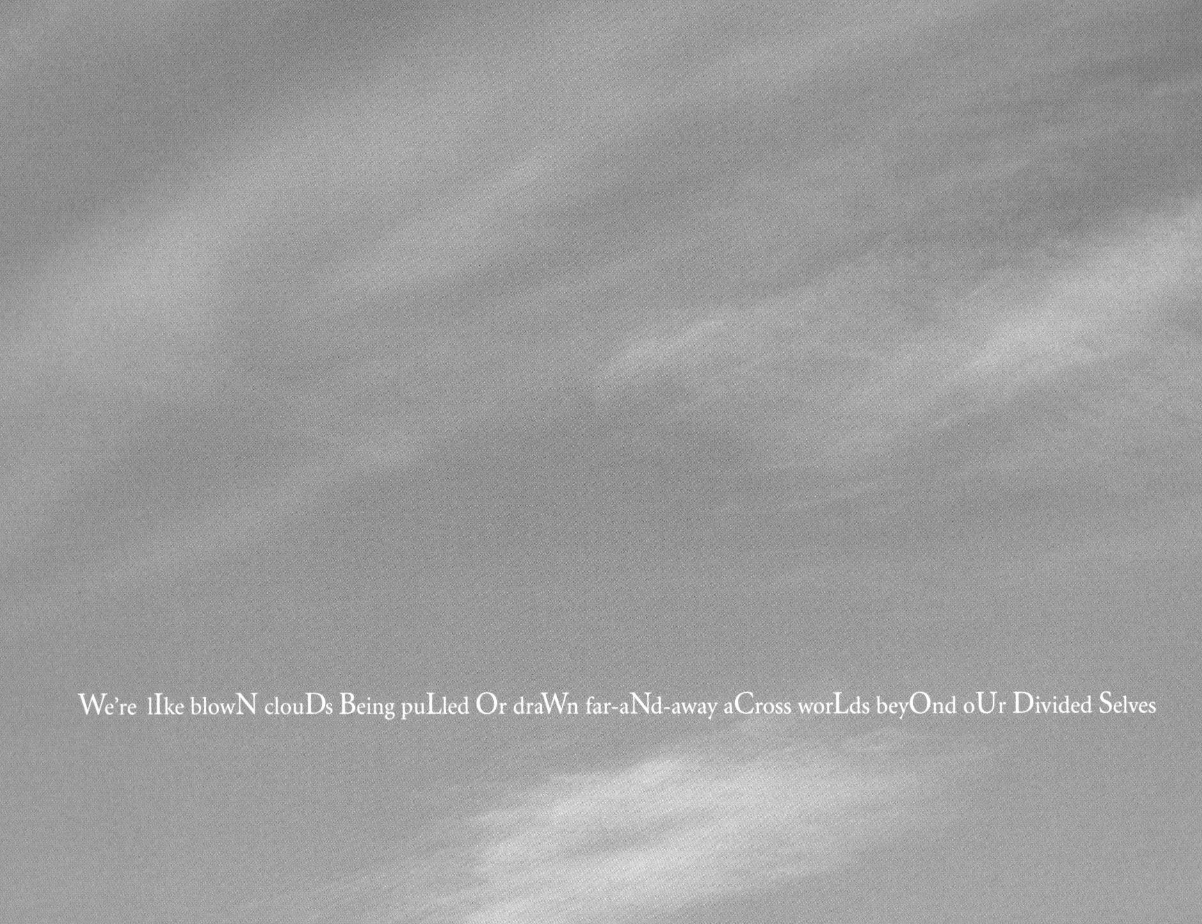
We're lIke blowN clouDs Being puLled Or draWn far-aNd-away aCross worLds beyOnd oUr Divided Selves

I want the water, I want the air

I began to collect wind blown clouds in the spring of 1999. The idea came to me when I reread the Japanese haiku poet Basho's *The Road to the Far North*, which describes a pilgrimage through the Shirakawa barrier and highlands, to the west coast and a view over the sea to the Isle of Sado. The image Basho used for his wanderlust is that of being *drawn like a wind blown cloud*; come spring we all know that feeling.

The project is an invitation to the world to gather an ongoing record of the sky. On any morning I may find a parcel on my doorstep filled with blue and white slides from River City or Punta del Este, Gwangju, or Kinloch Rannoch. Each one is carefully catalogued and then stored in *THE ARCHIVE OF WIND BLOWN CLOUDS*. The invitation—numerous postcard and Web versions of which have been published—introduced me to people like Amparo, who was eight years old when she first wrote to me, and whose cloud photographs are among the most beautiful of all those I have received.

The archive appears as if it were a scientific project, gathering data that awaits classification; but there will be no such rational arrangement, for its purpose is poetic, utopian. This is a collaboration with amateurs and enthusiasts. My principle has been to accept every contribution to the archive and, in the published selections I list all of the authors, without identifying individual clouds by photographer, place, or date. Here then are clouds—small ones, fluffy ones, filets, broad bands, and golden towers—nothing more; here is a reminder that, despite political borders and human conflicts, we live under the same sky.

This is the second collection of clouds to appear. I would like to thank Rizzoli and all of the contributors for making this generous selection possible. The book is a gentle read and an instrument of transformation. It is also an invitation to send me some of your skies, for this is a project without end.

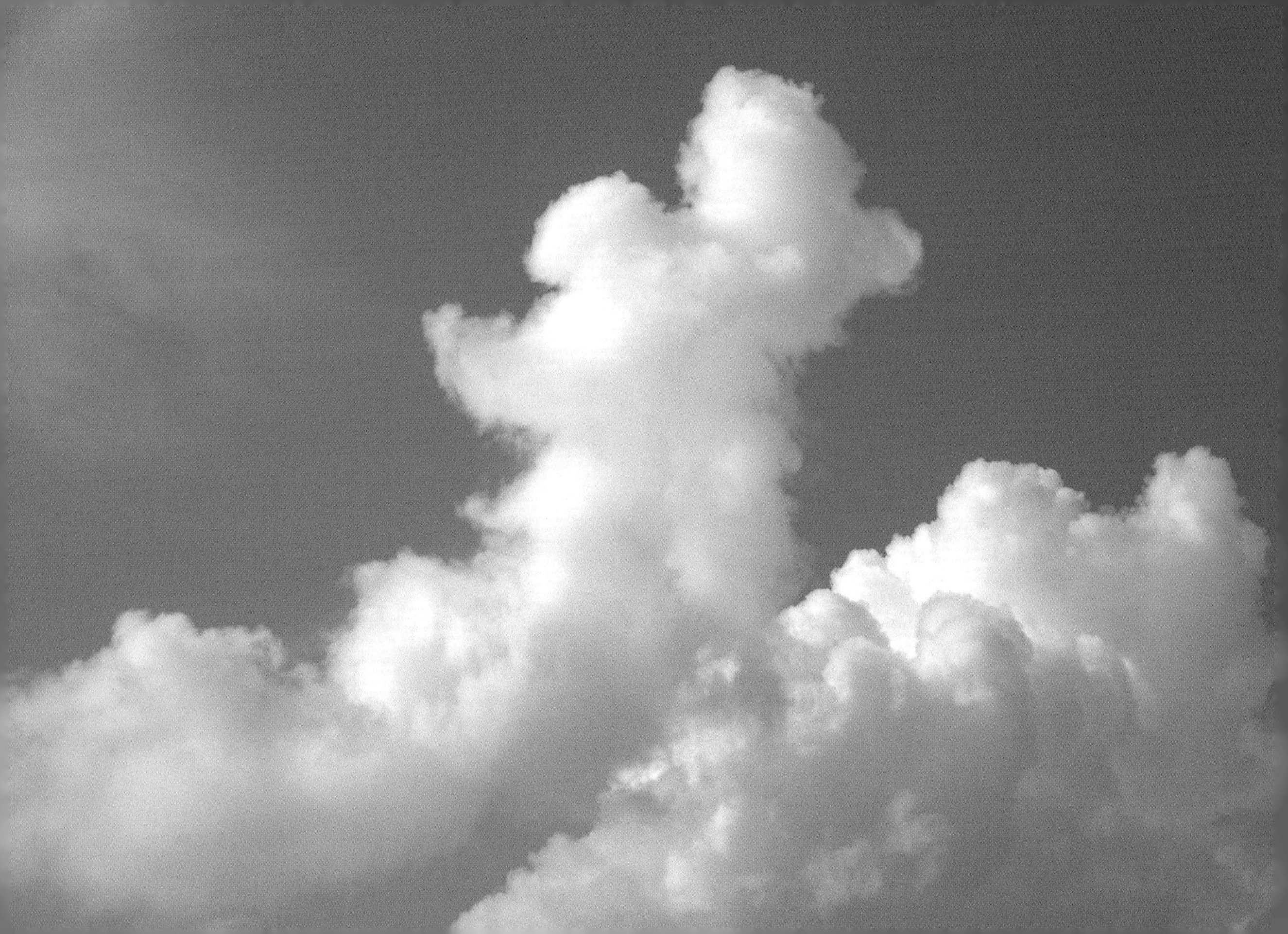

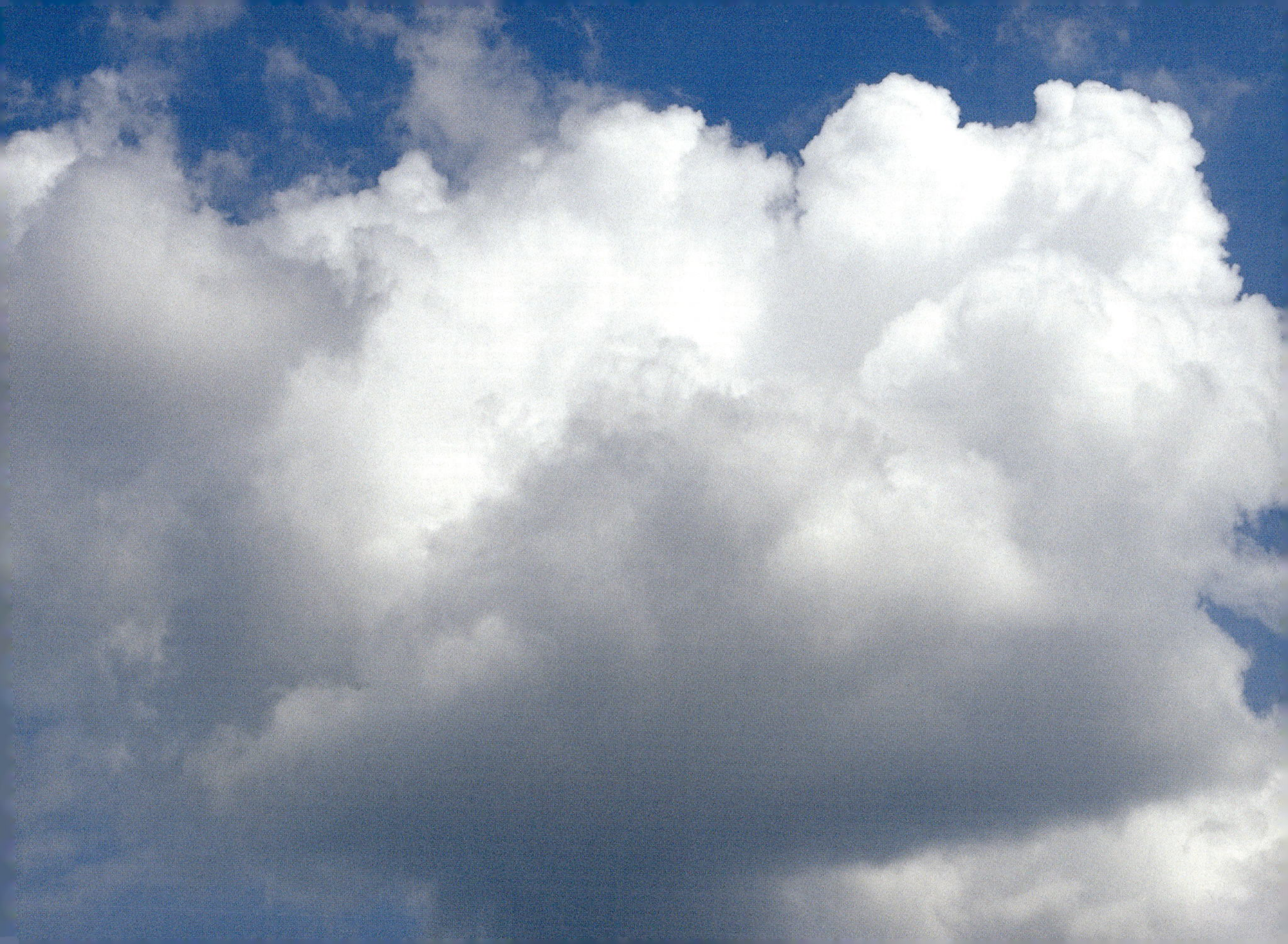

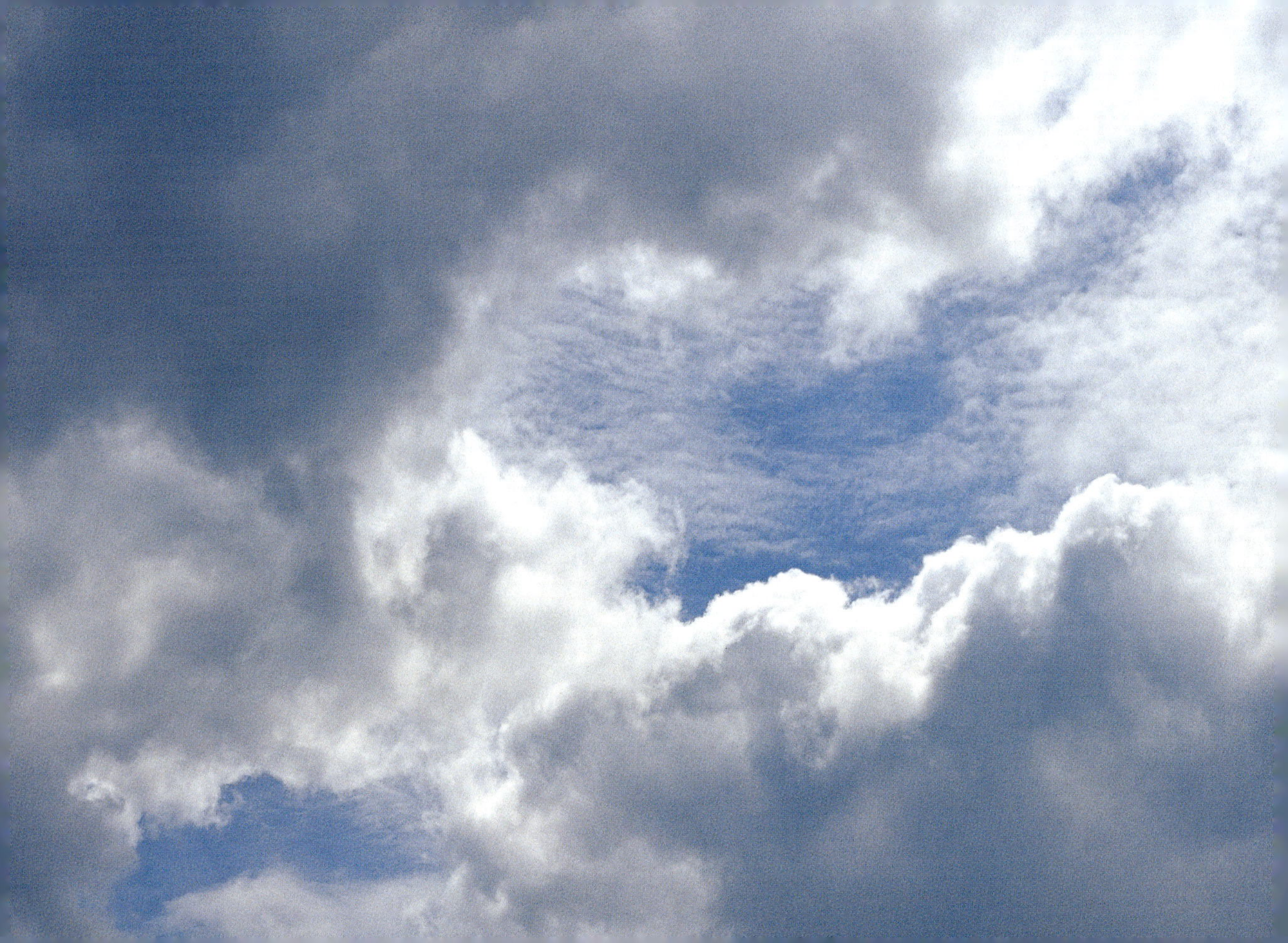

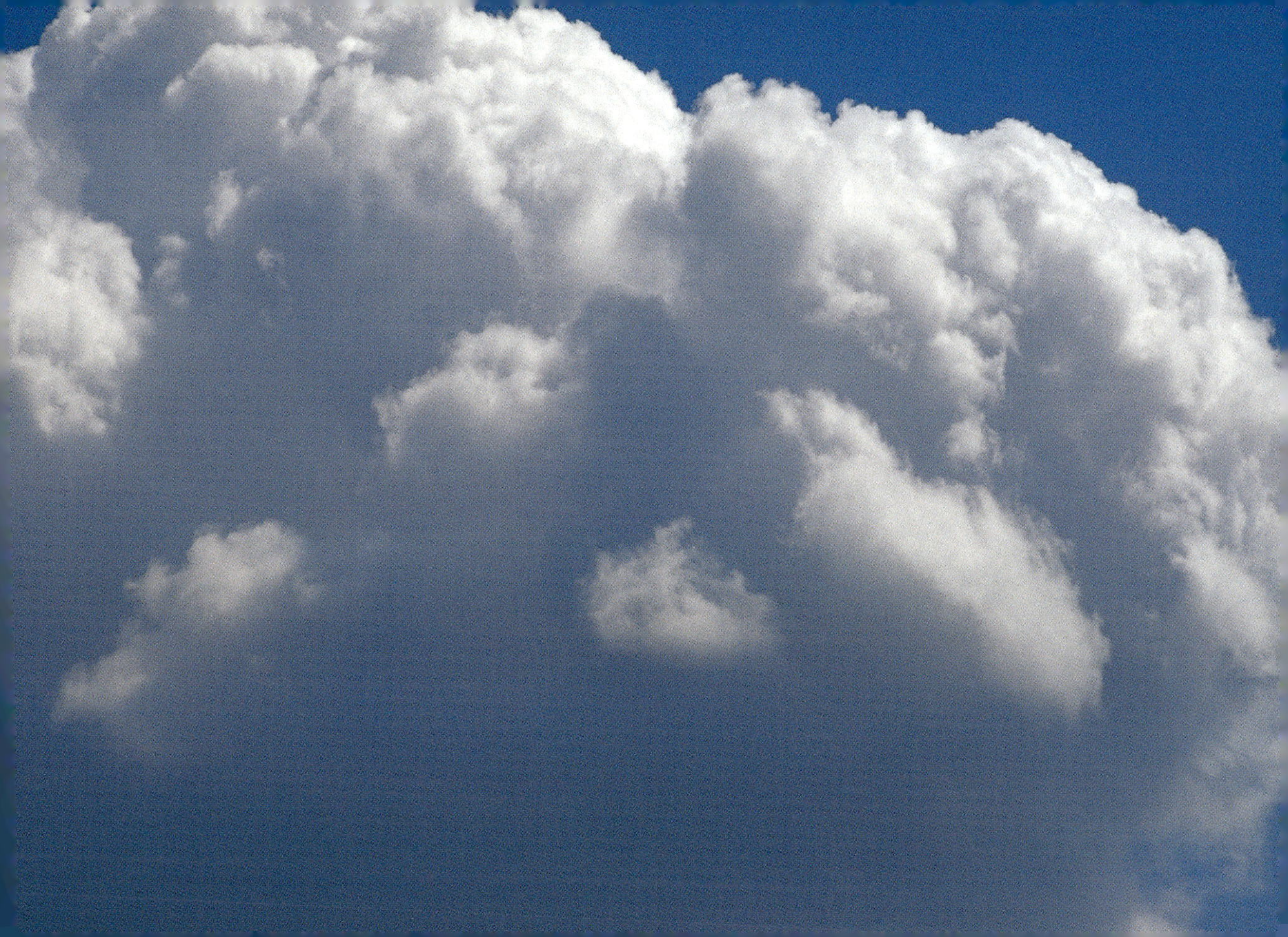

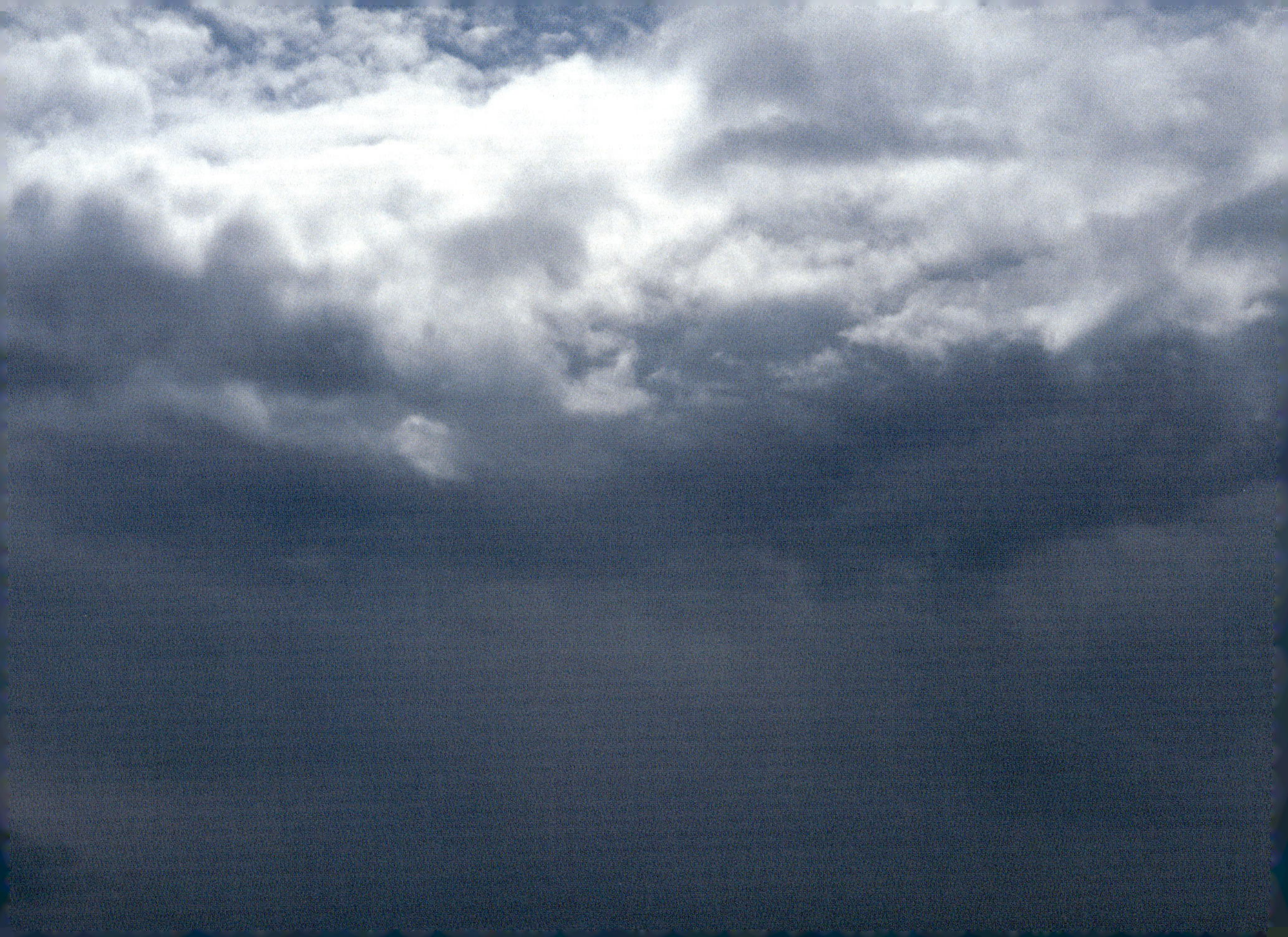

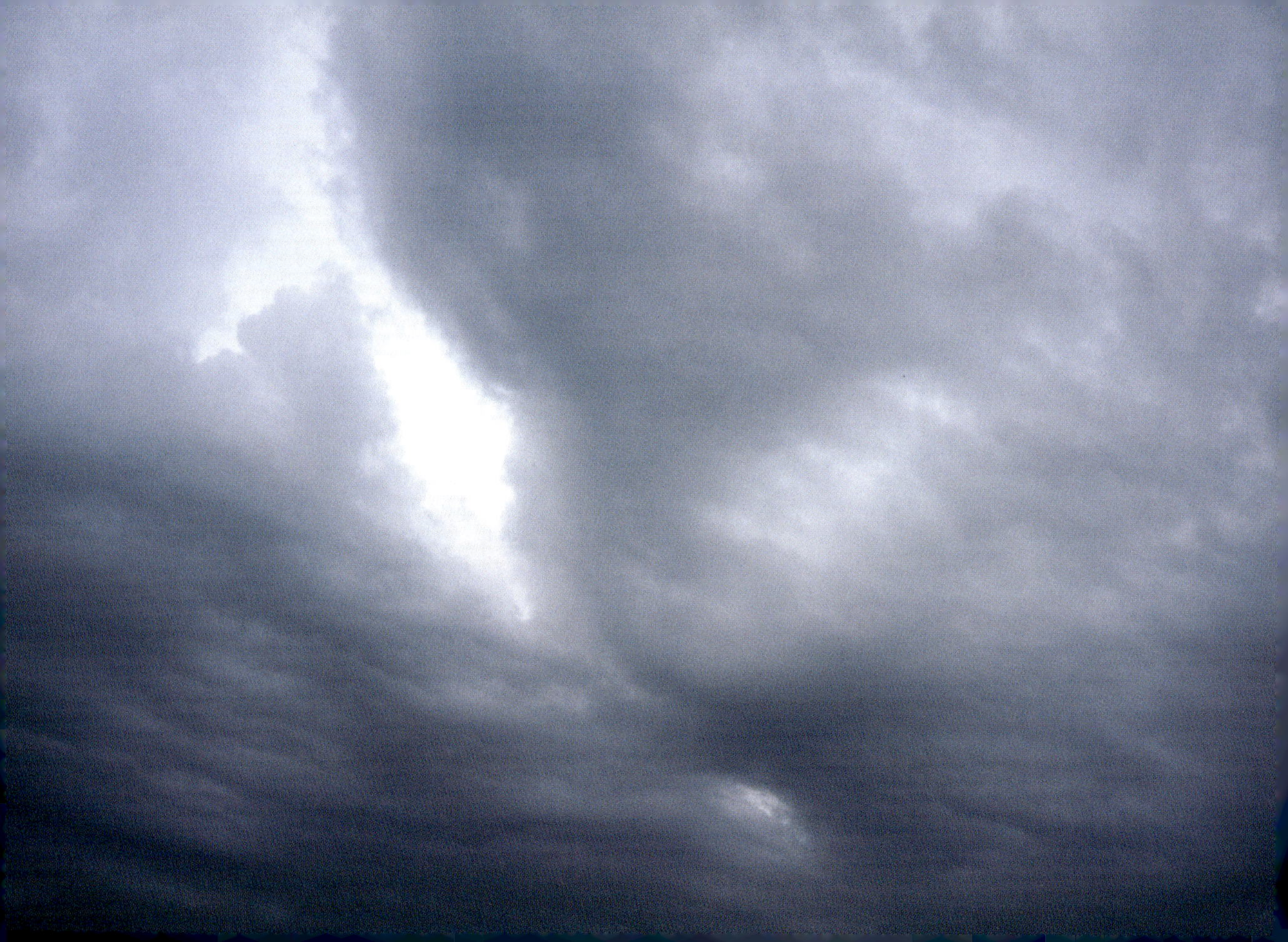

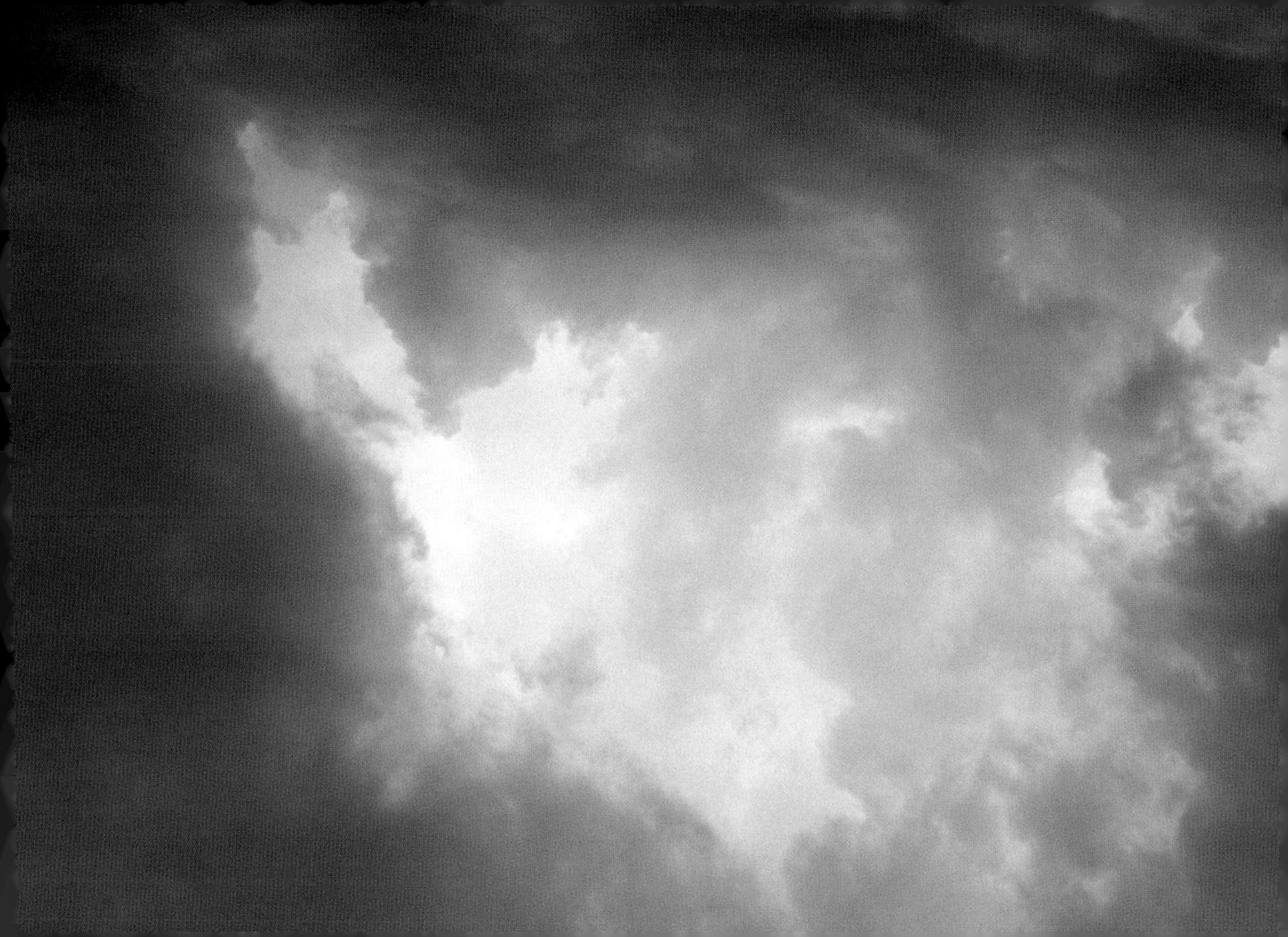

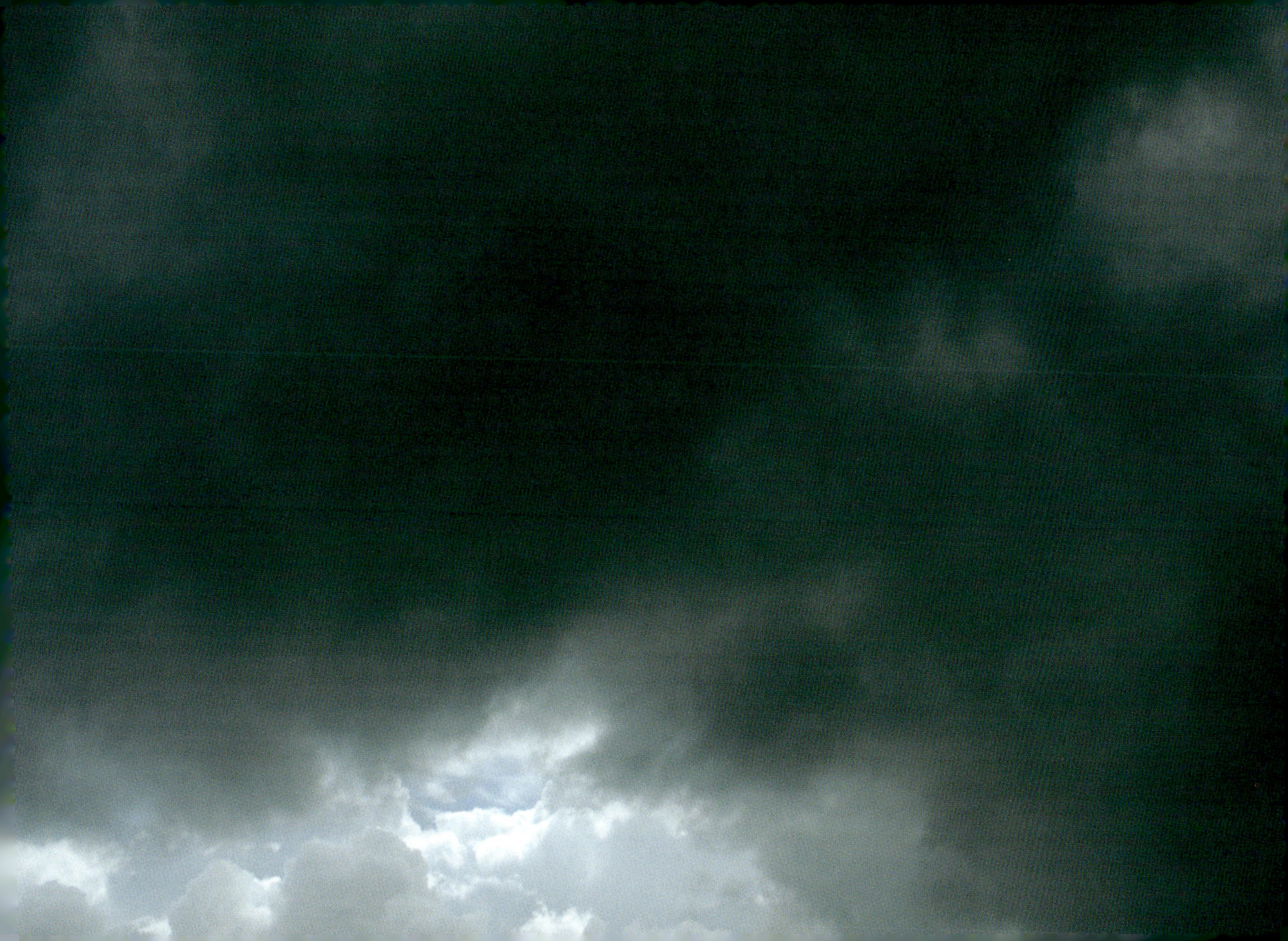

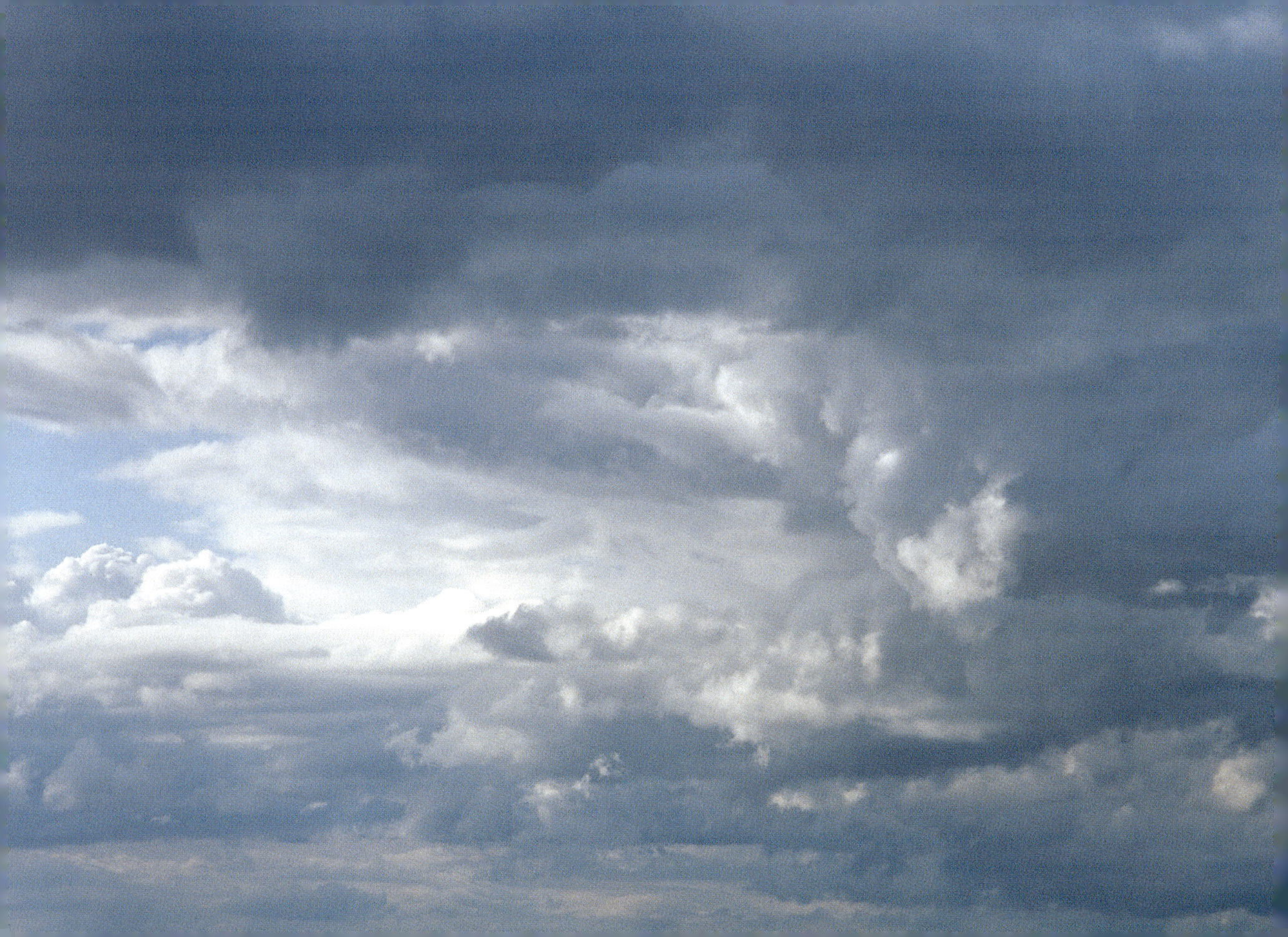

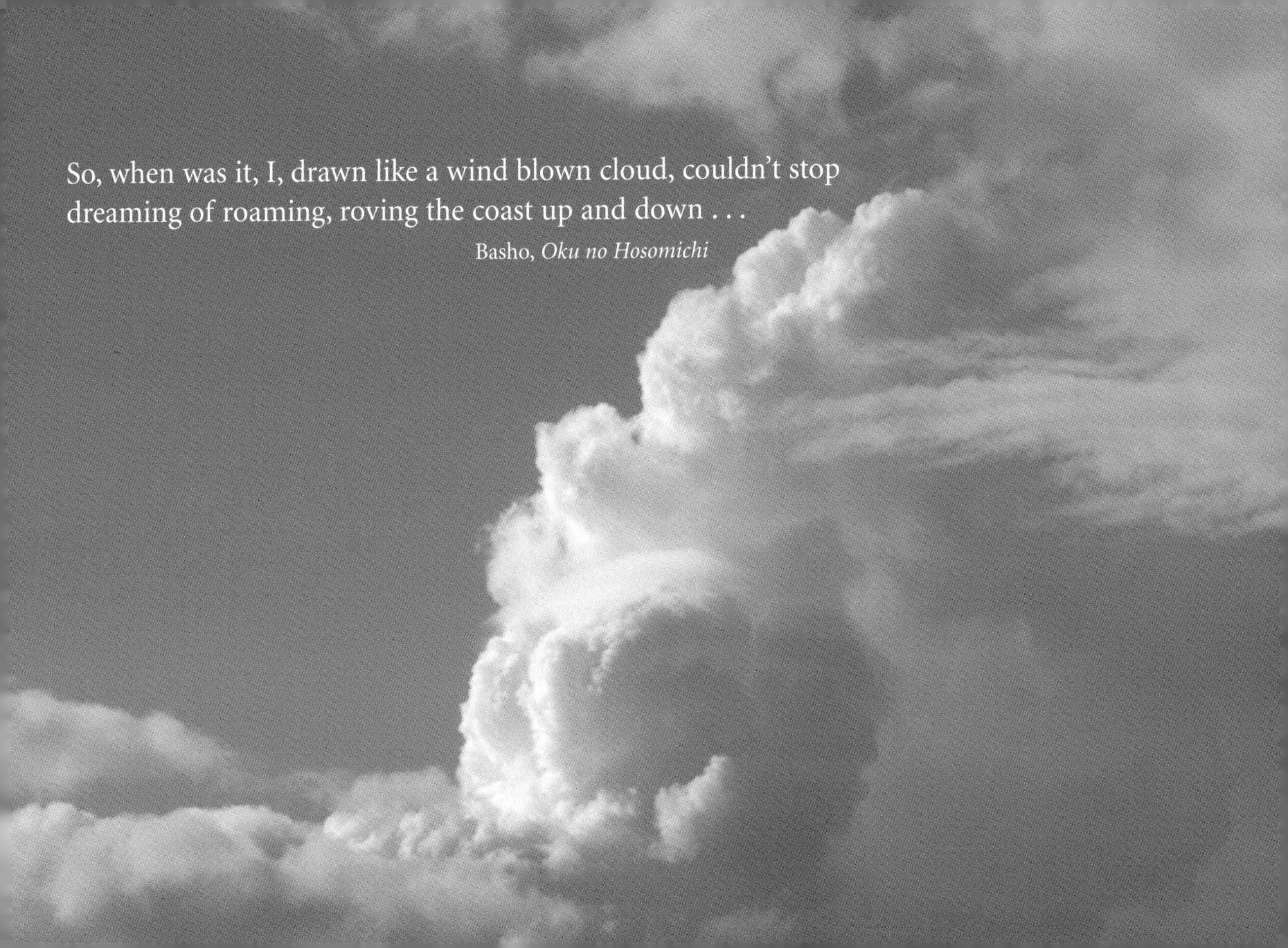

So, when was it, I, drawn like a wind blown cloud, couldn't stop
dreaming of roaming, roving the coast up and down . . .

Basho, *Oku no Hosomichi*

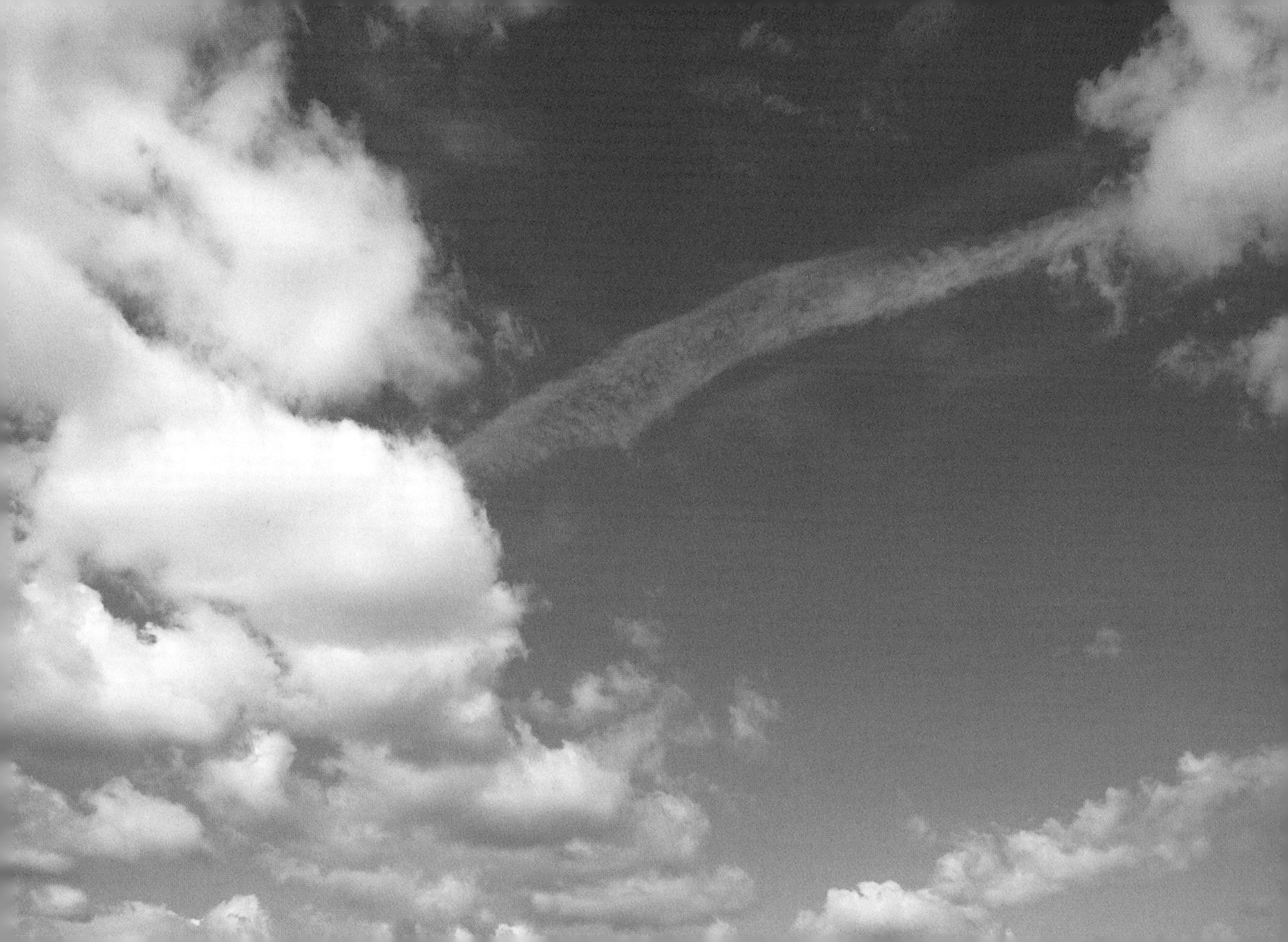

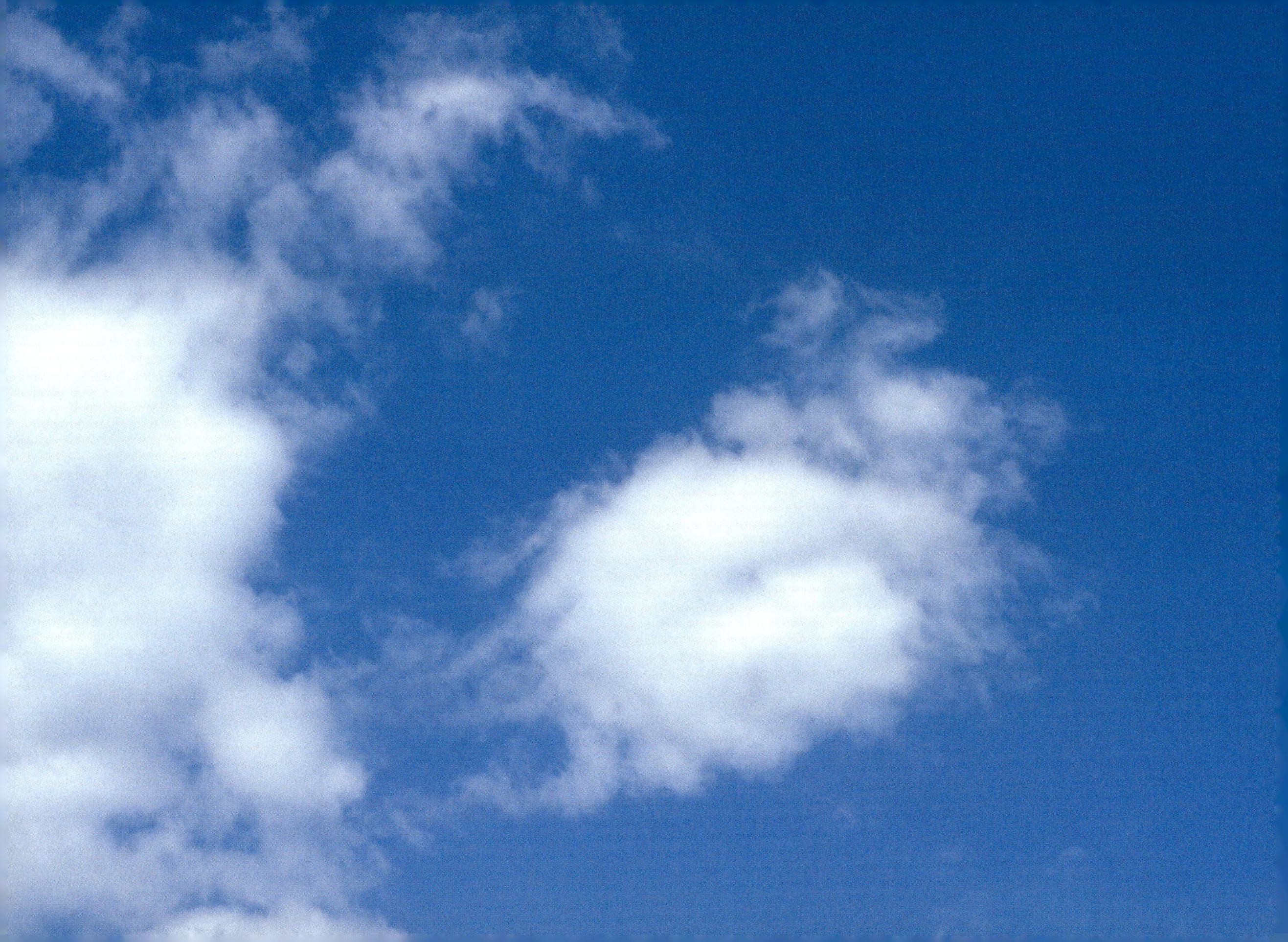

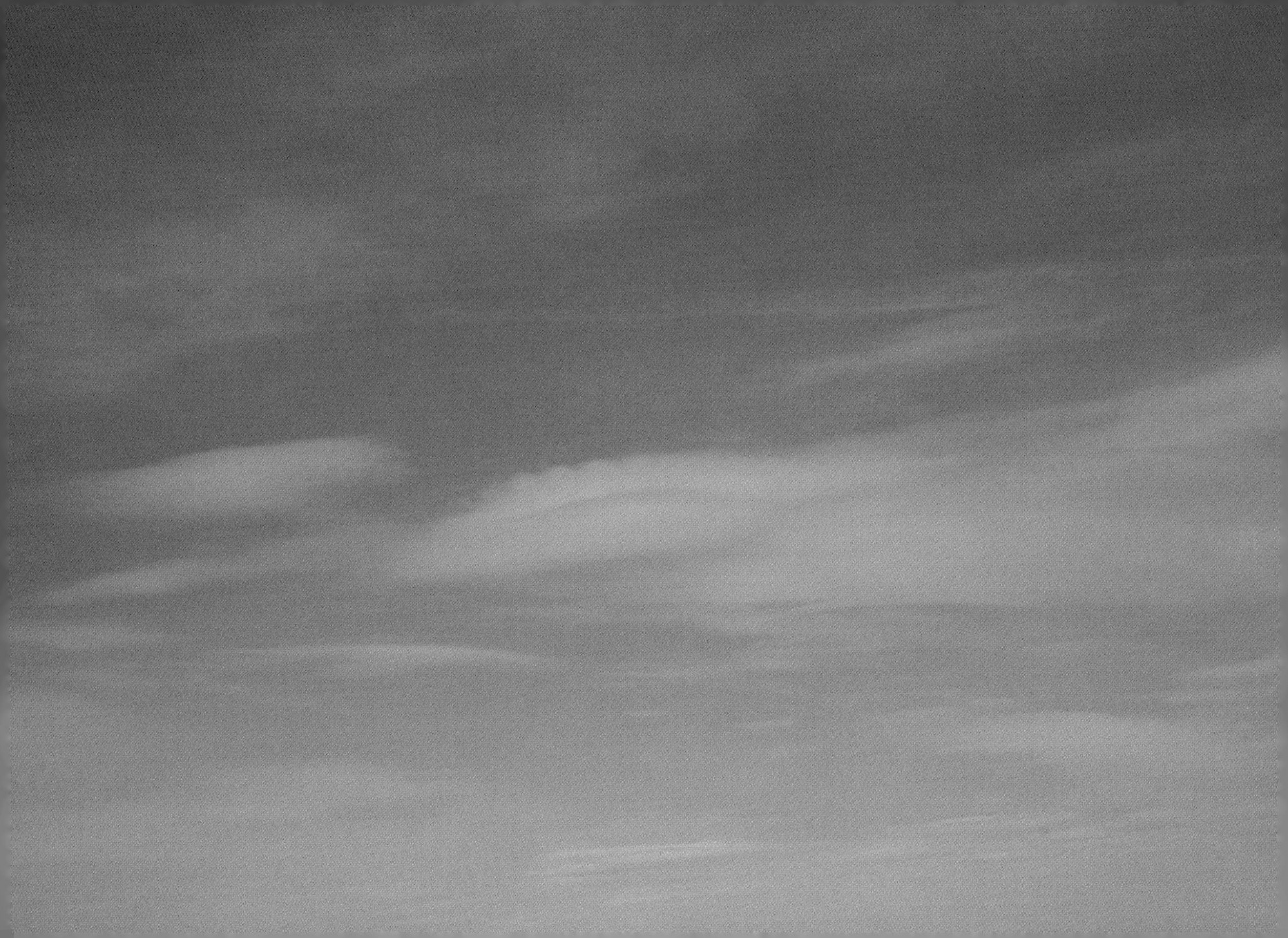

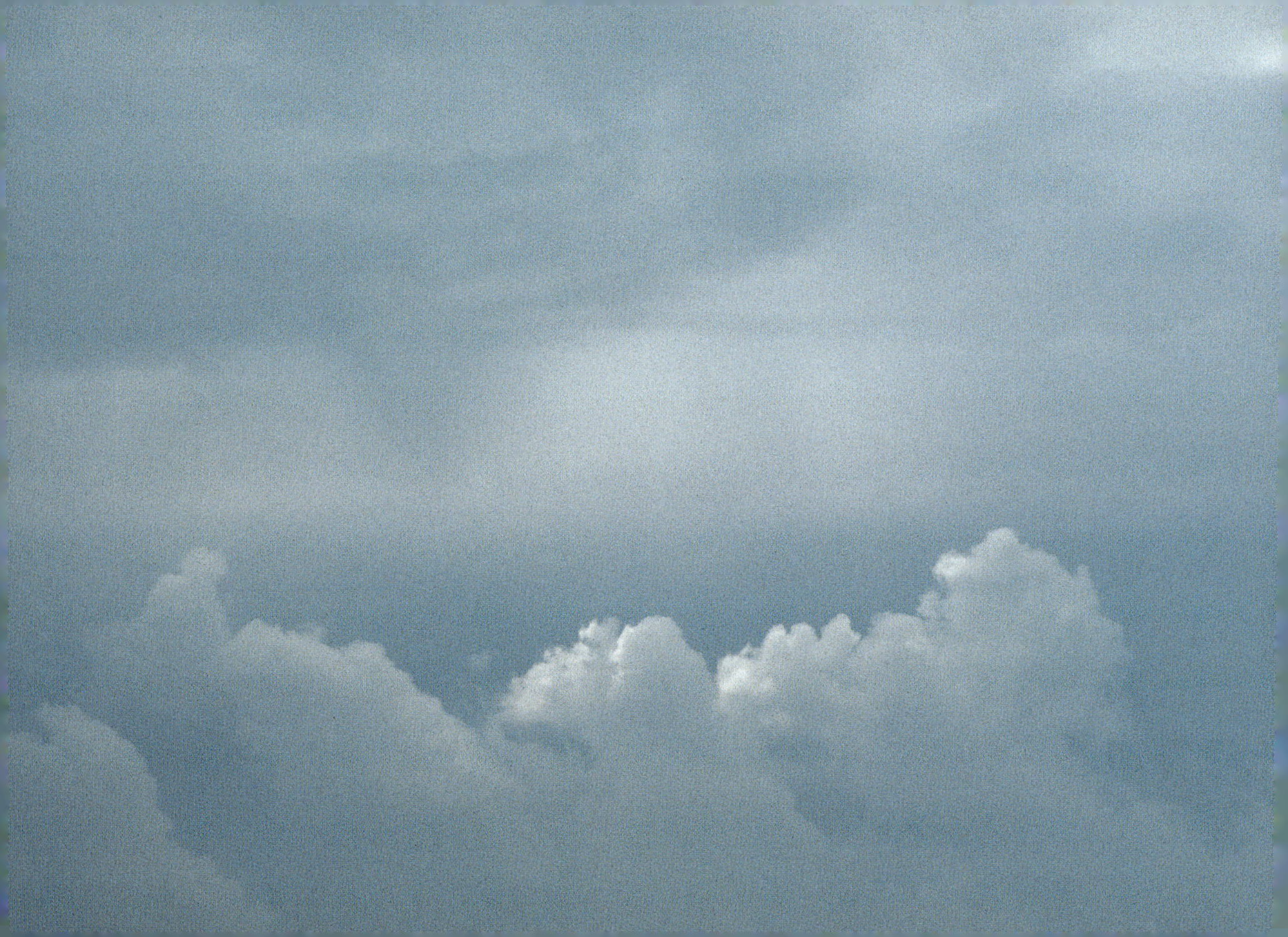

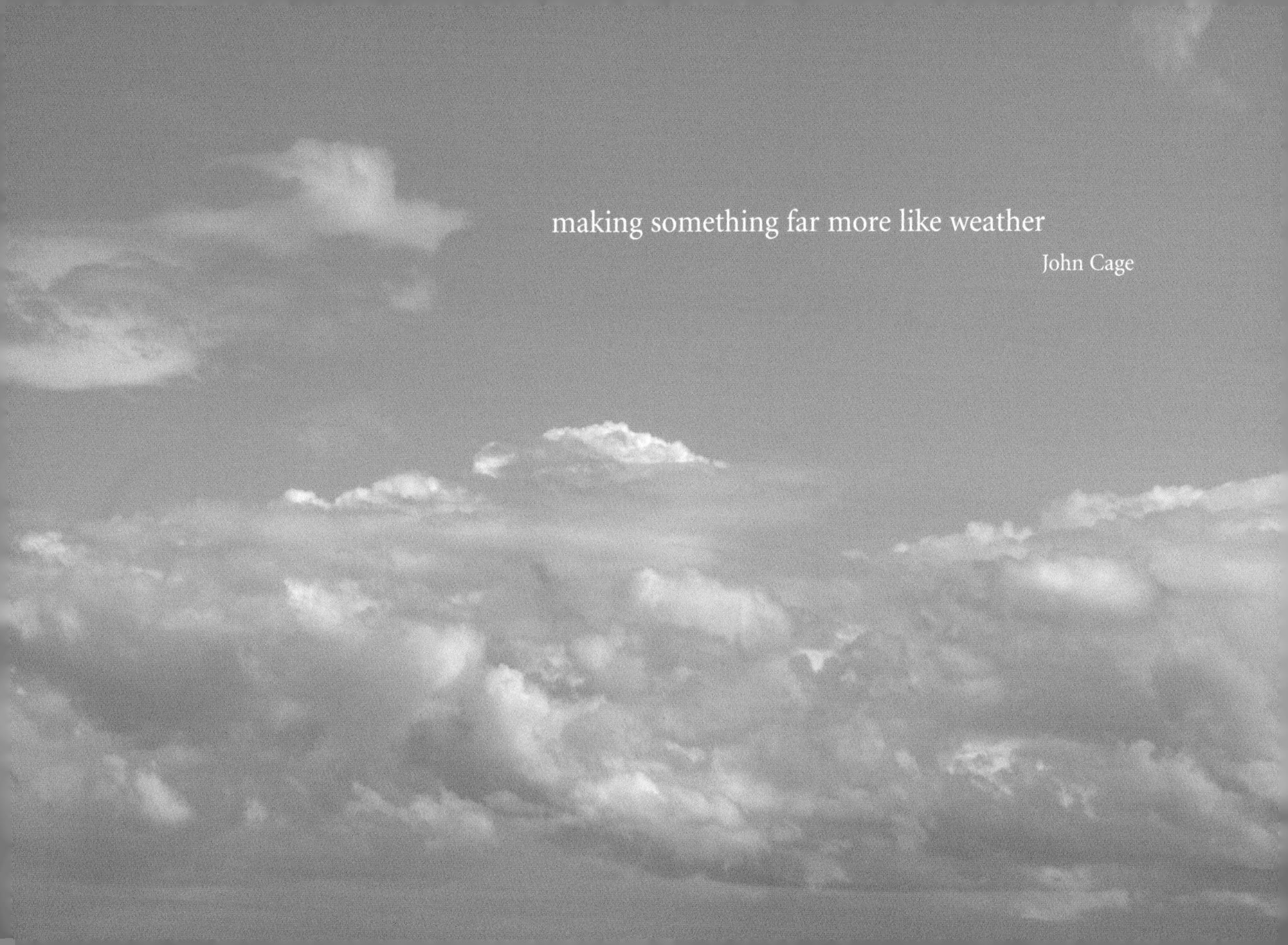

making something far more like weather

John Cage

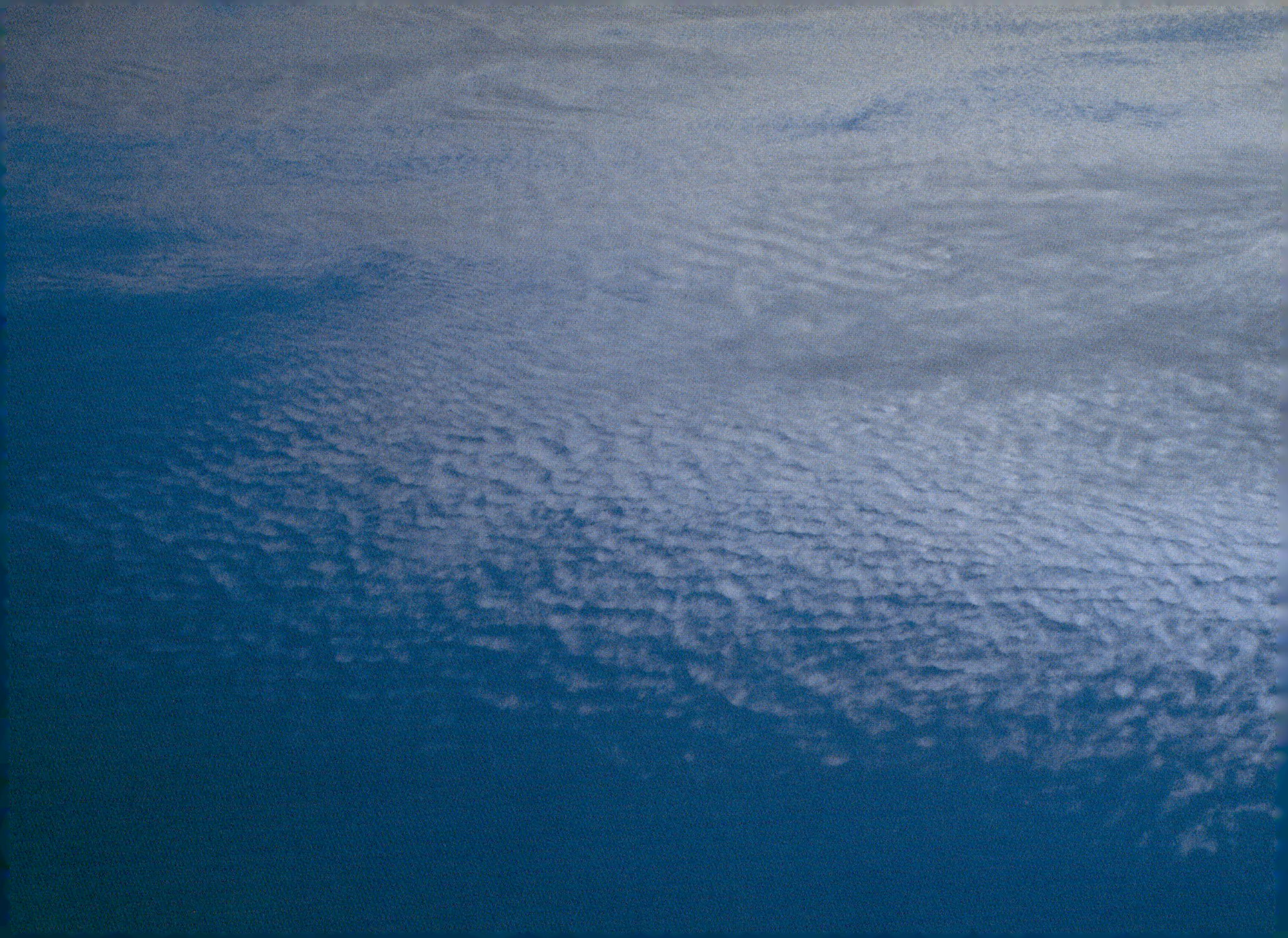

far clouds
white sails
crowding south

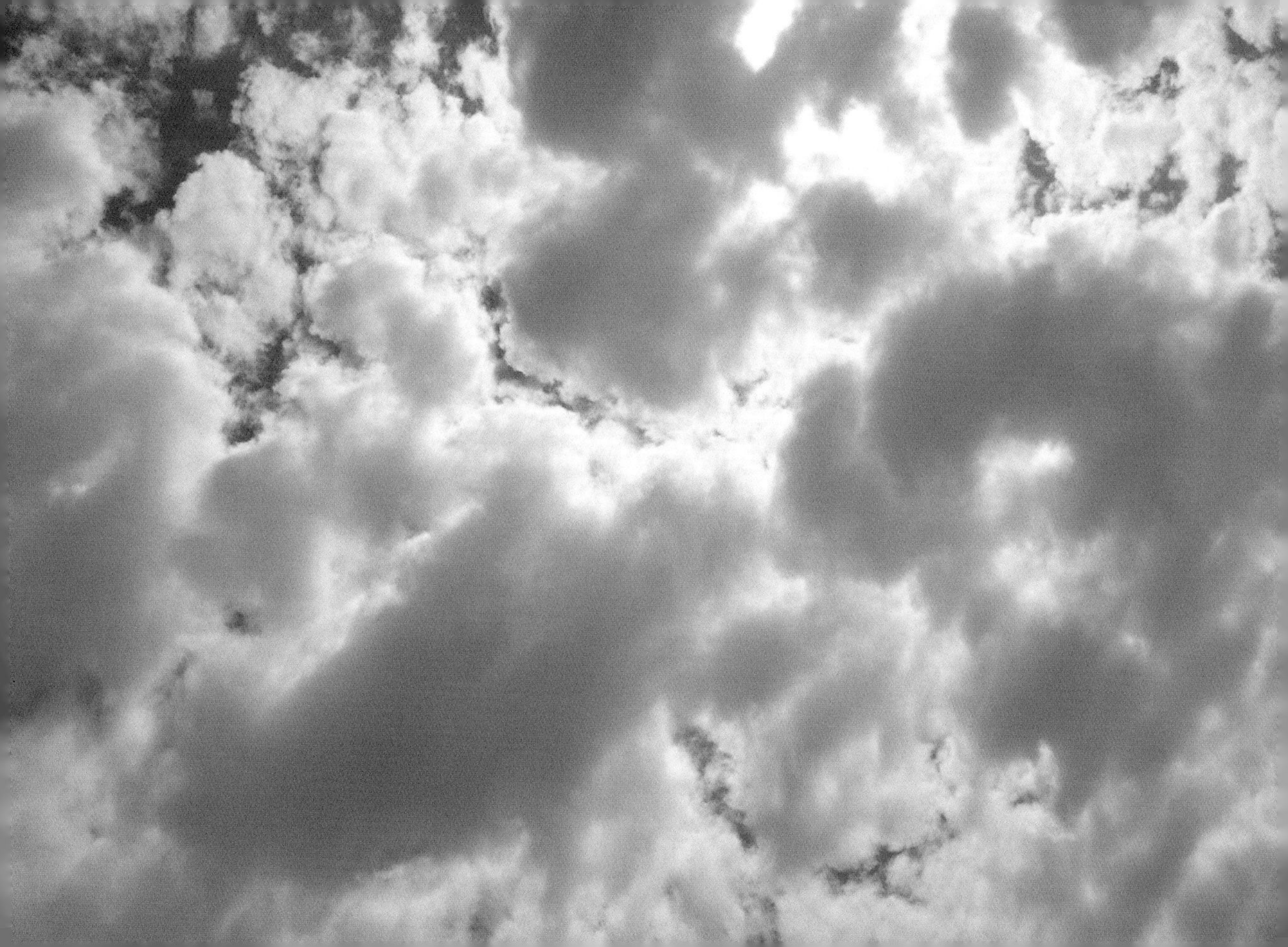

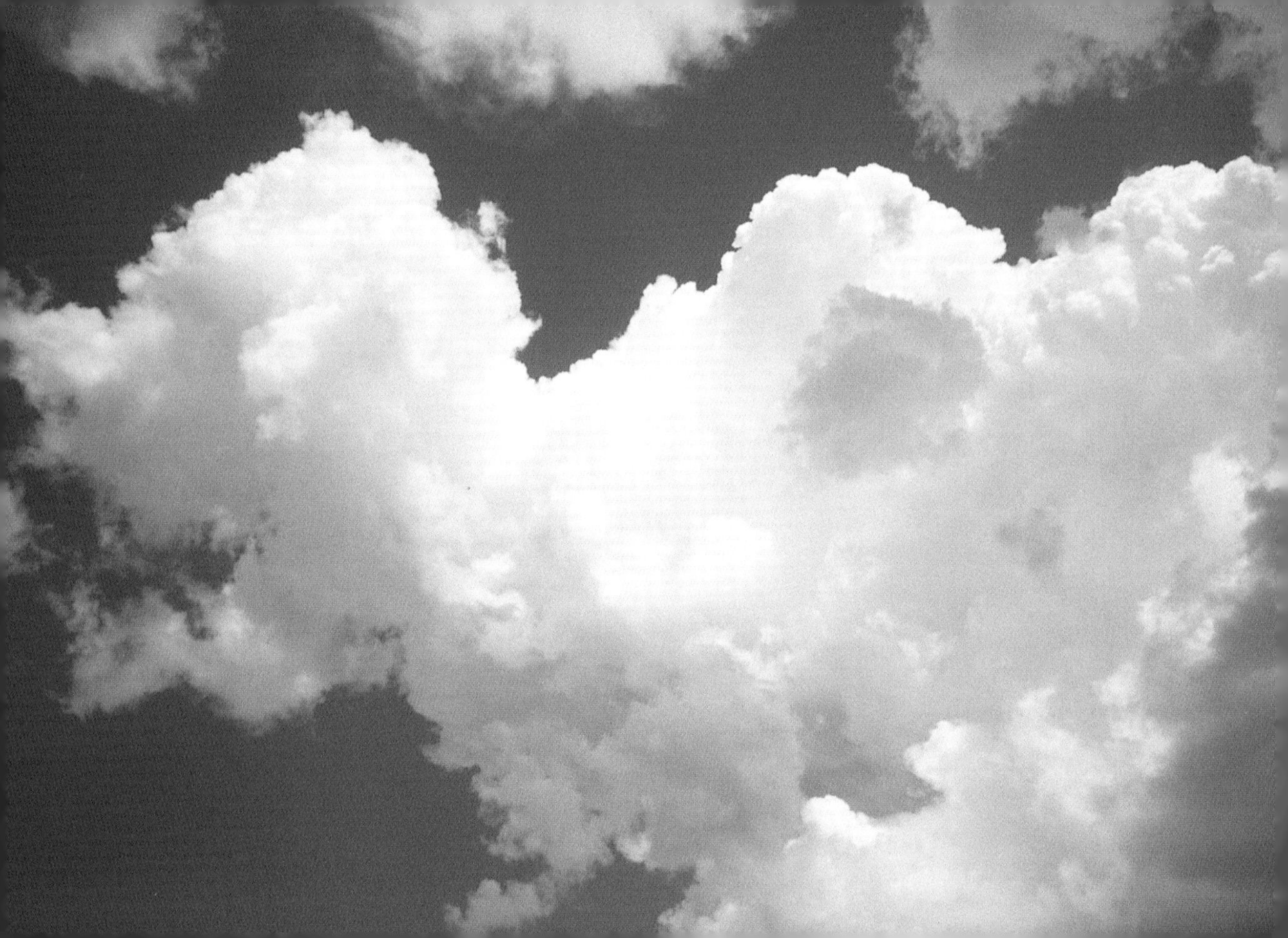

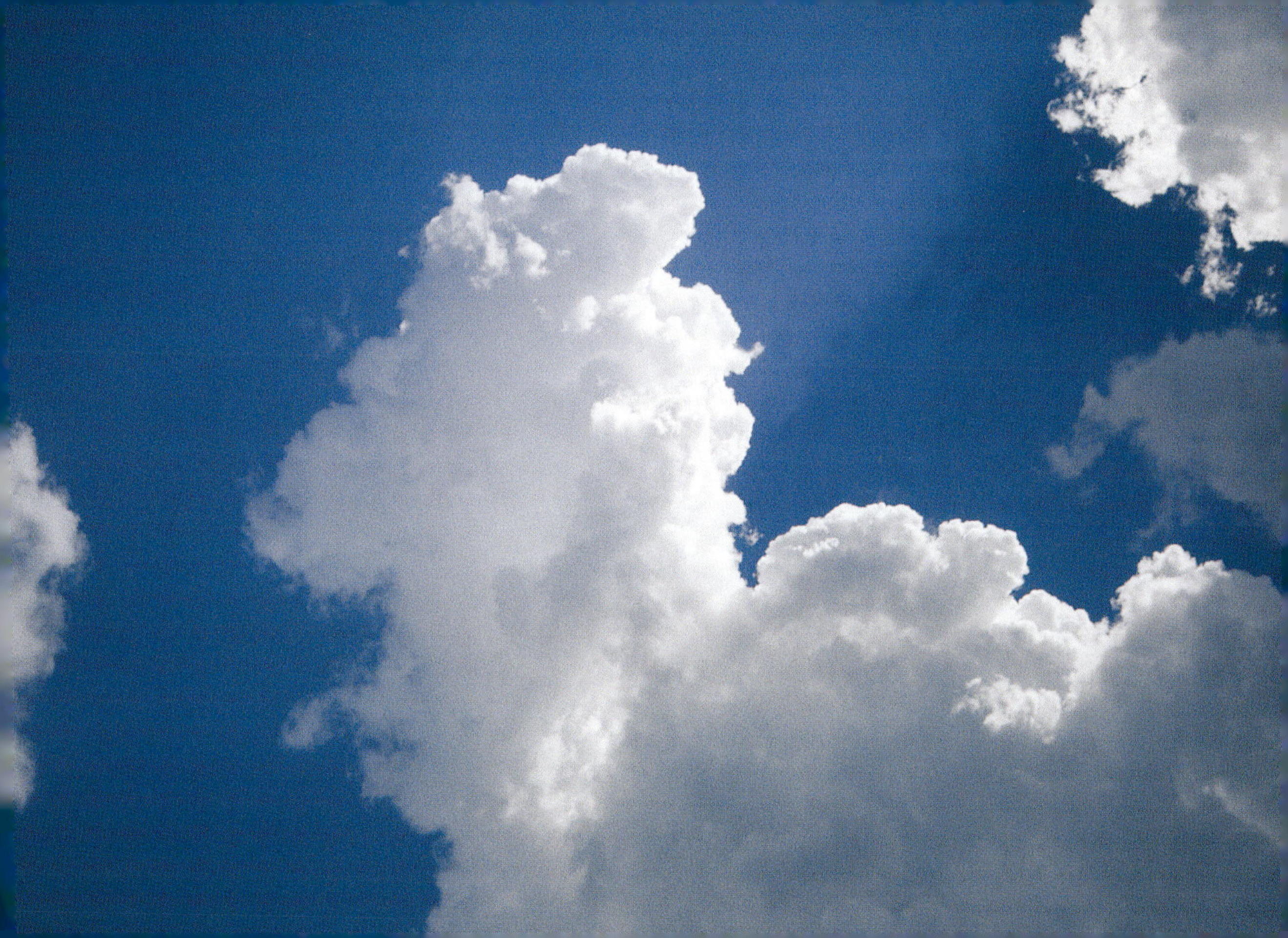

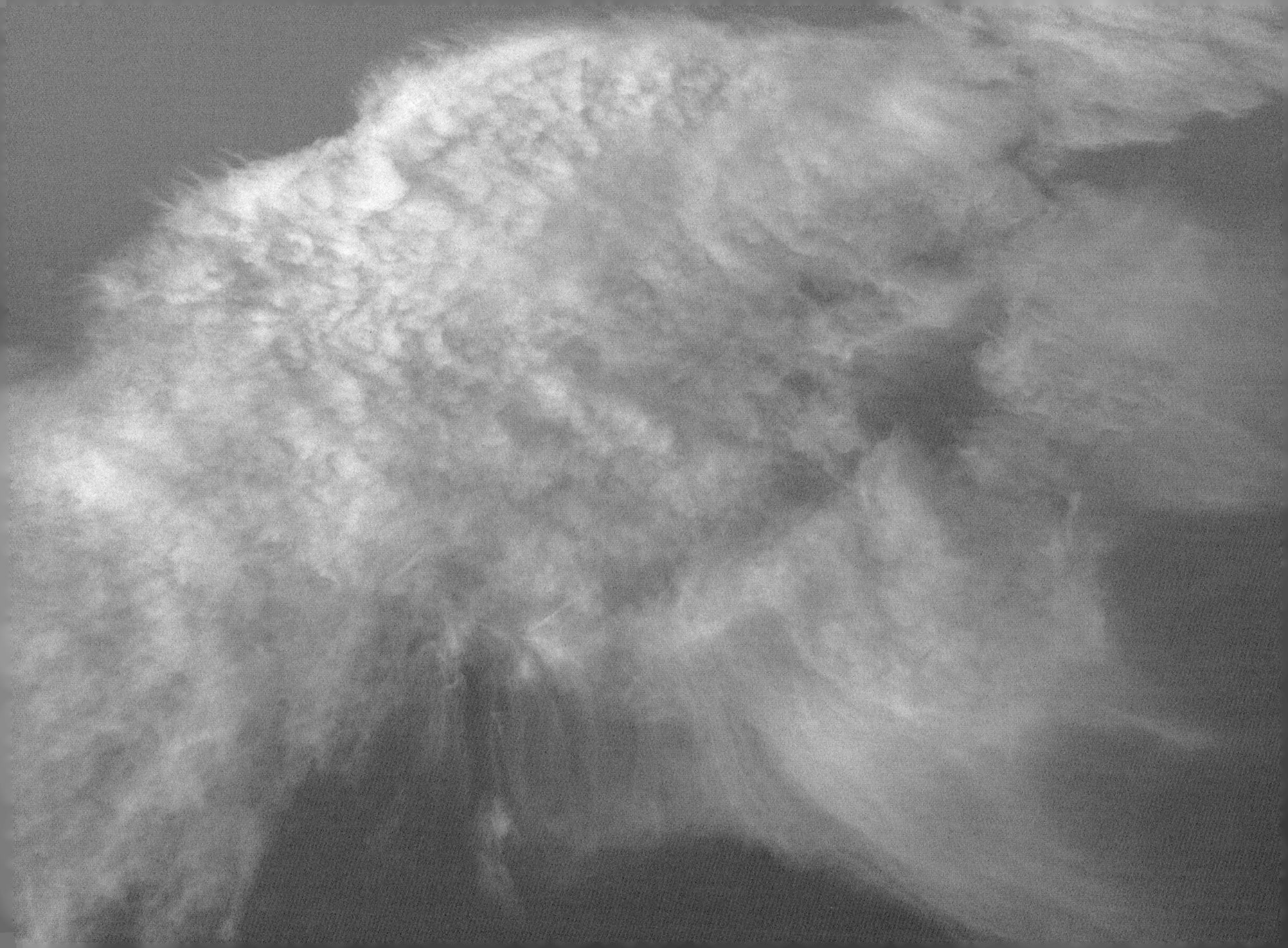

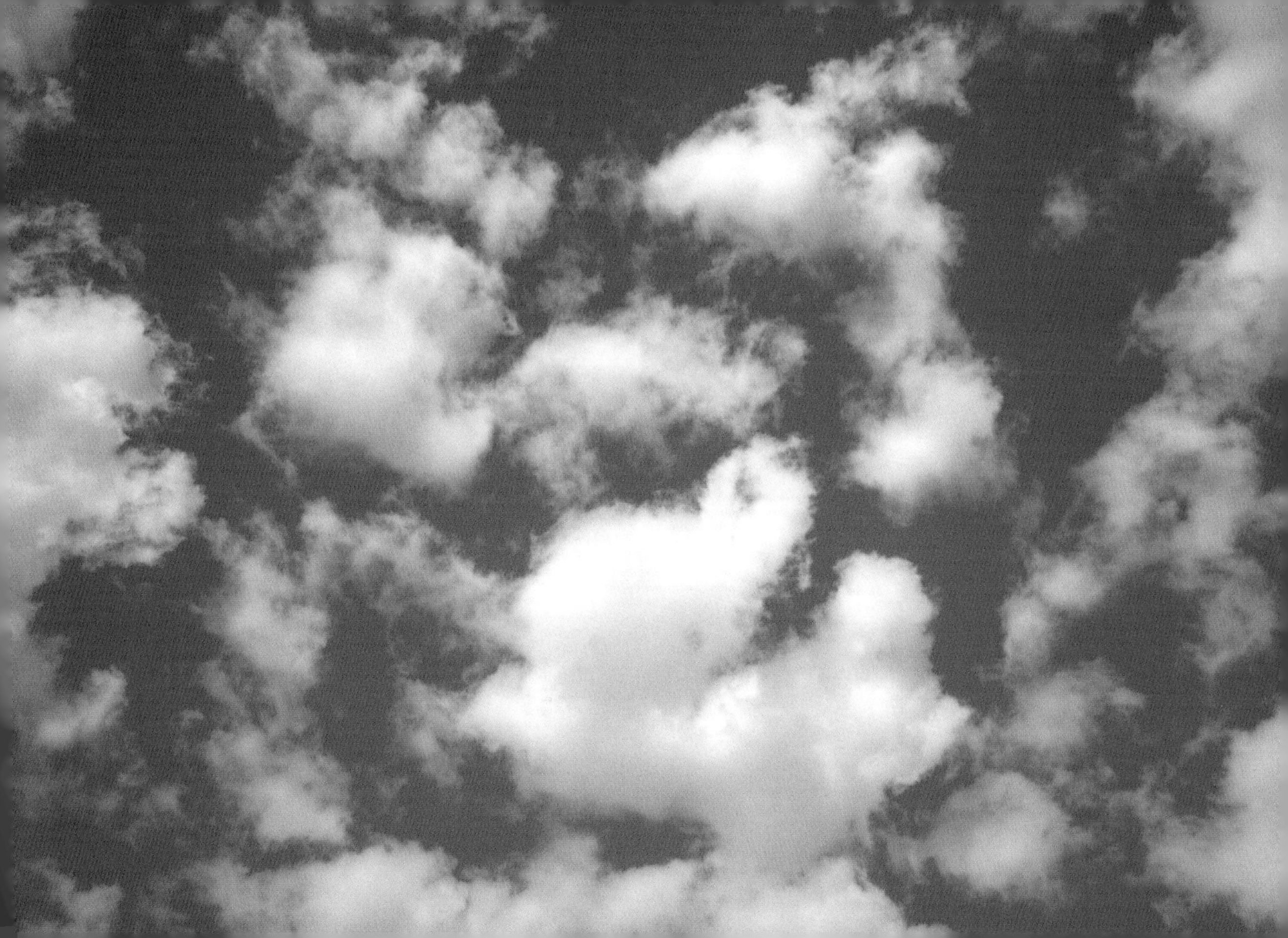

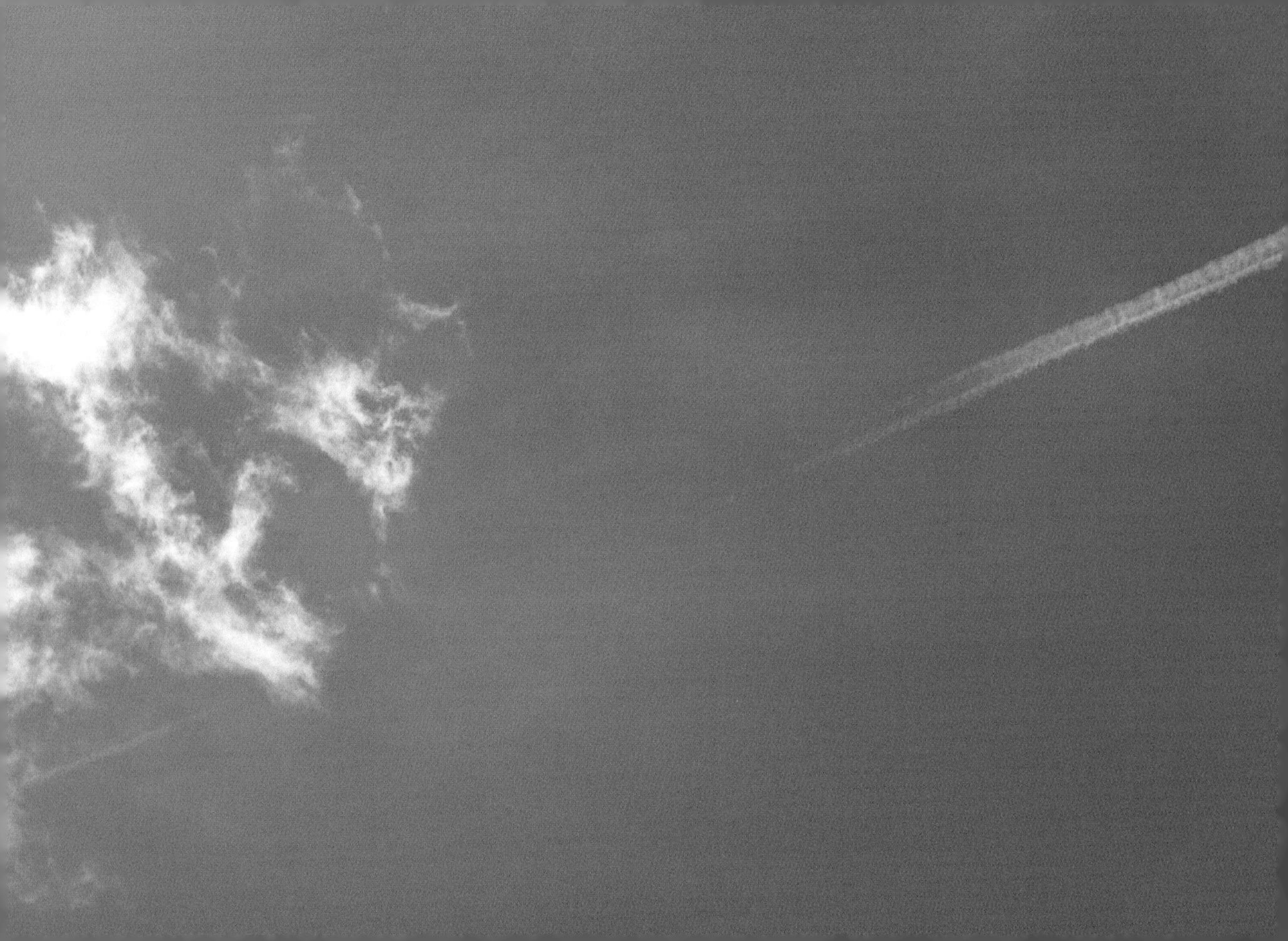

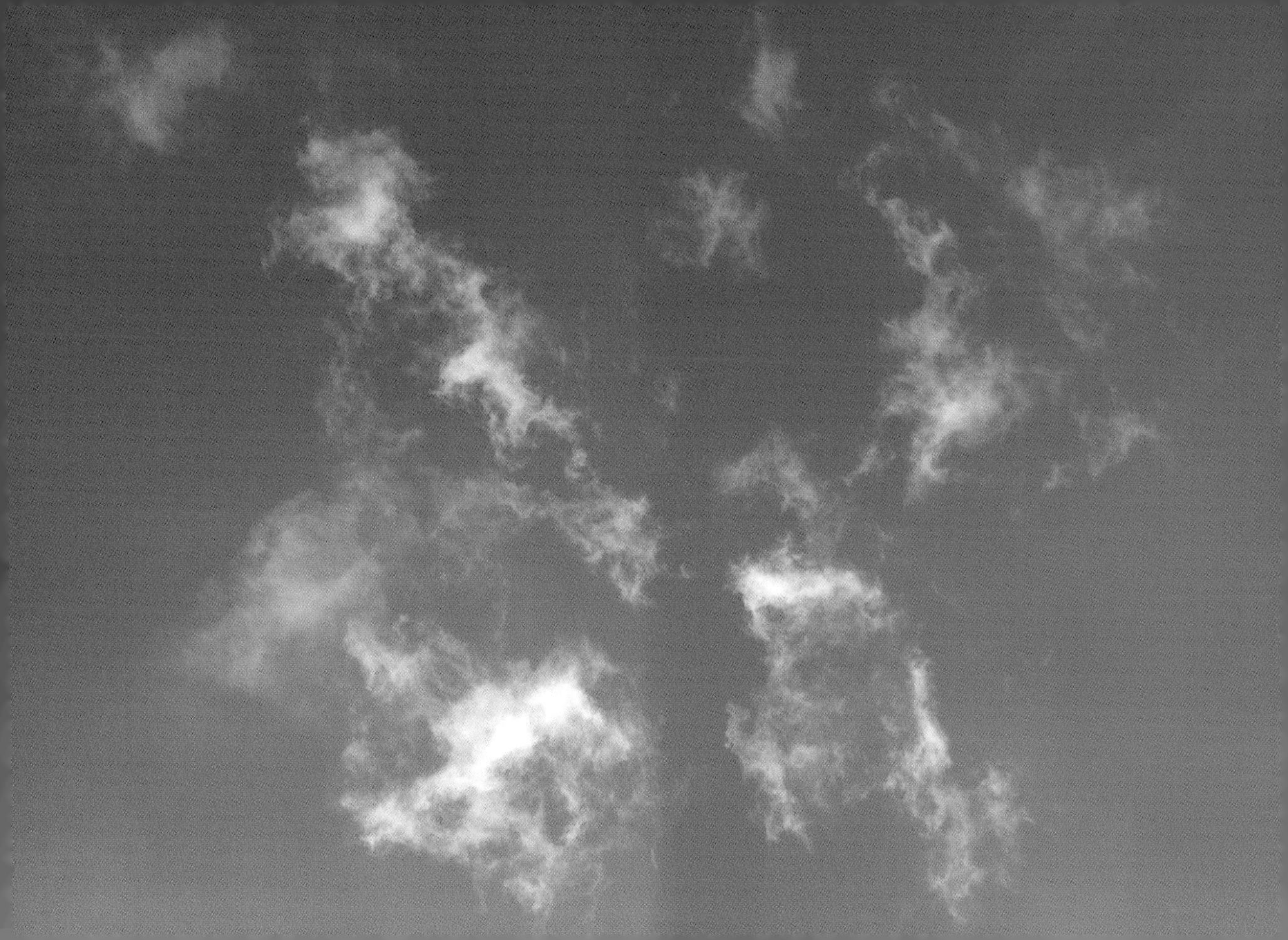

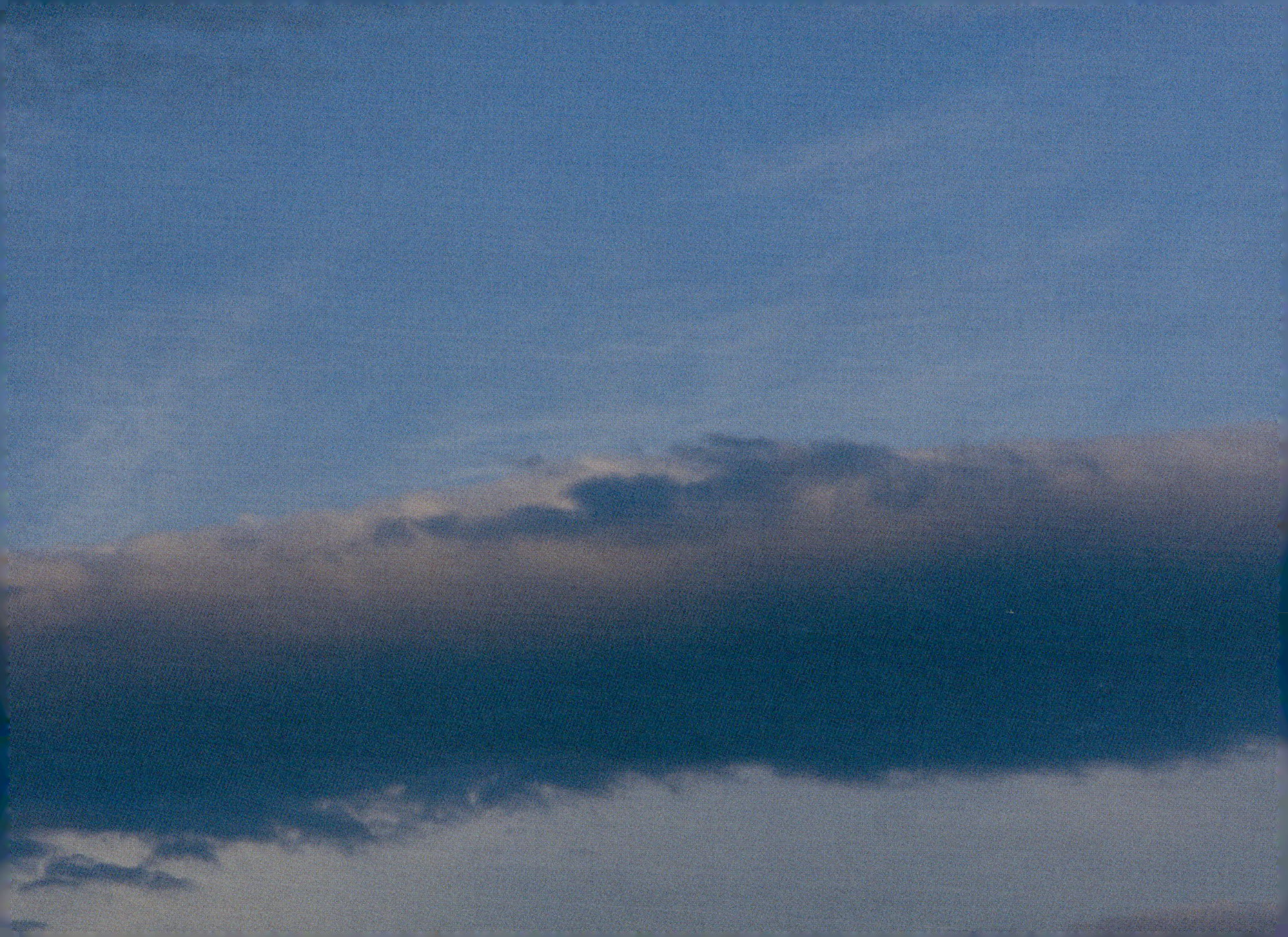

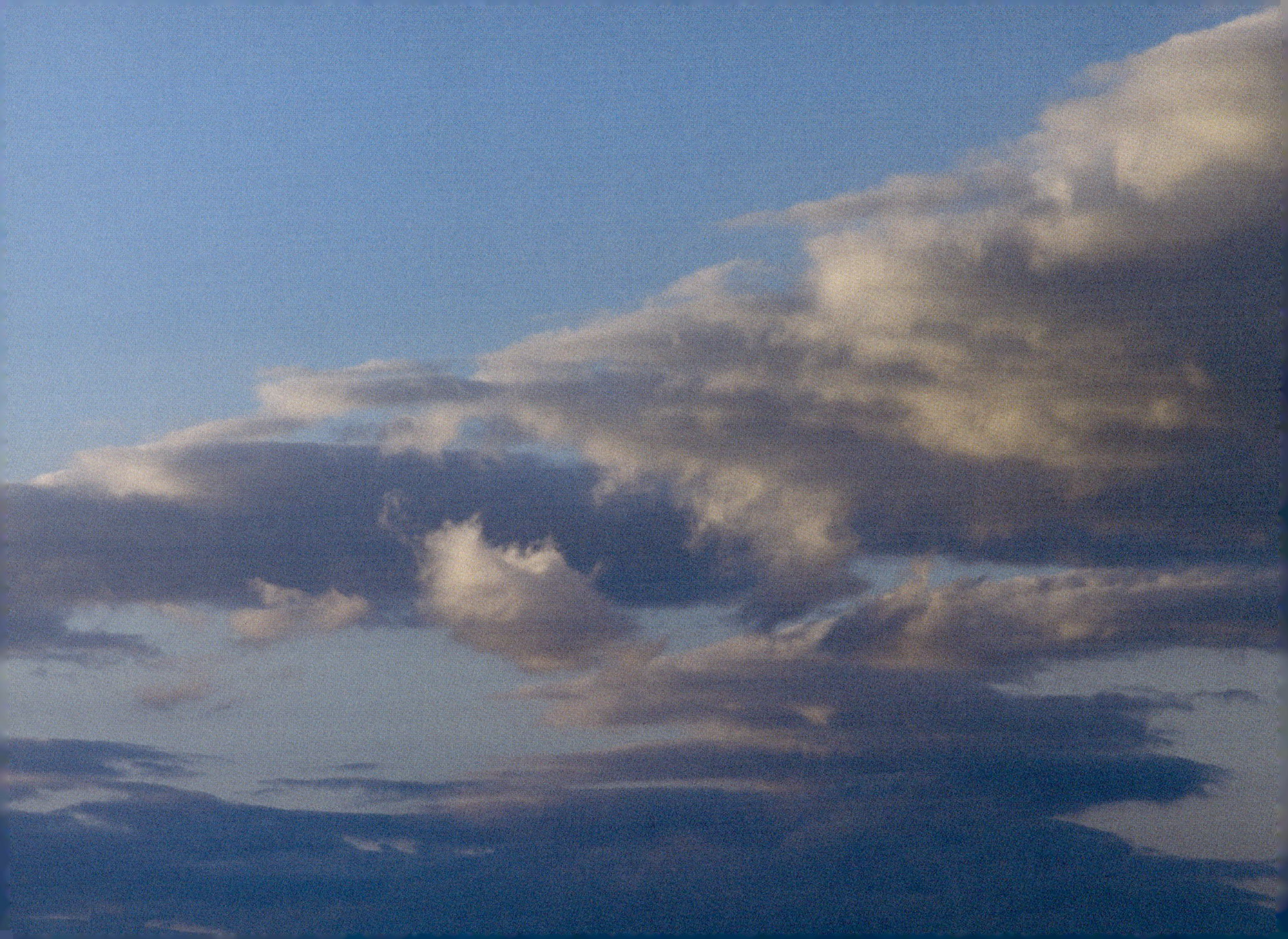

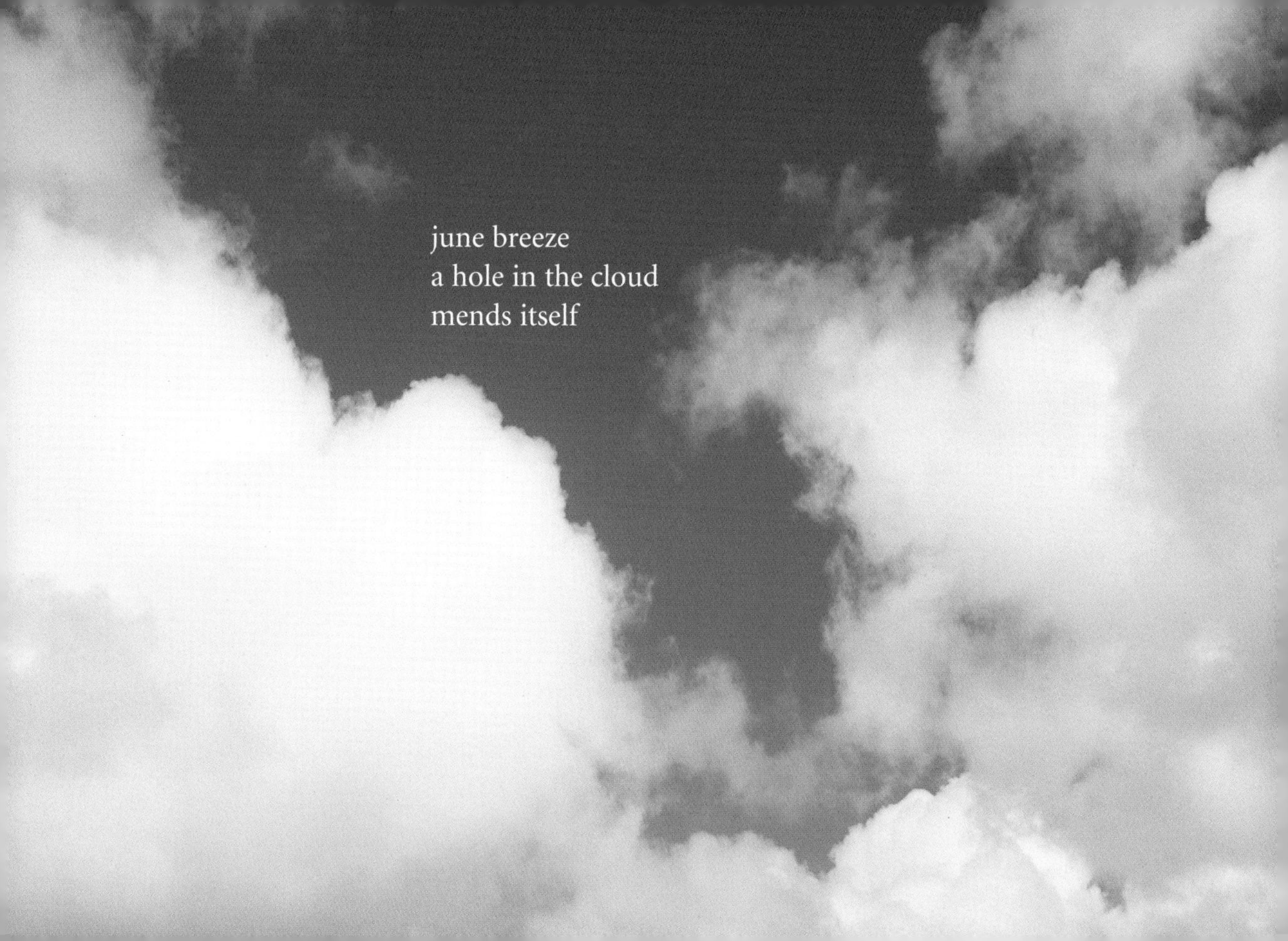

june breeze
a hole in the cloud
mends itself

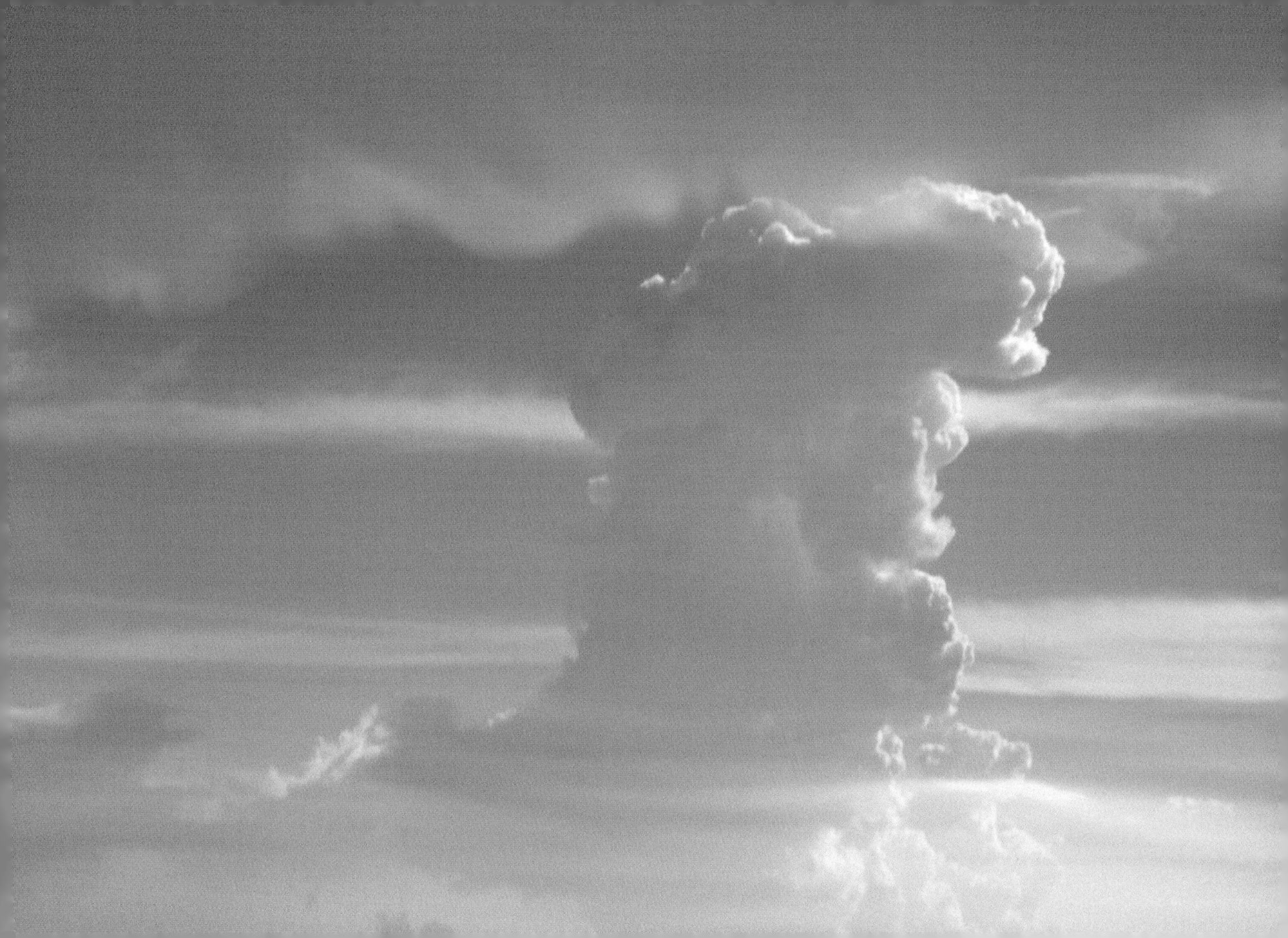

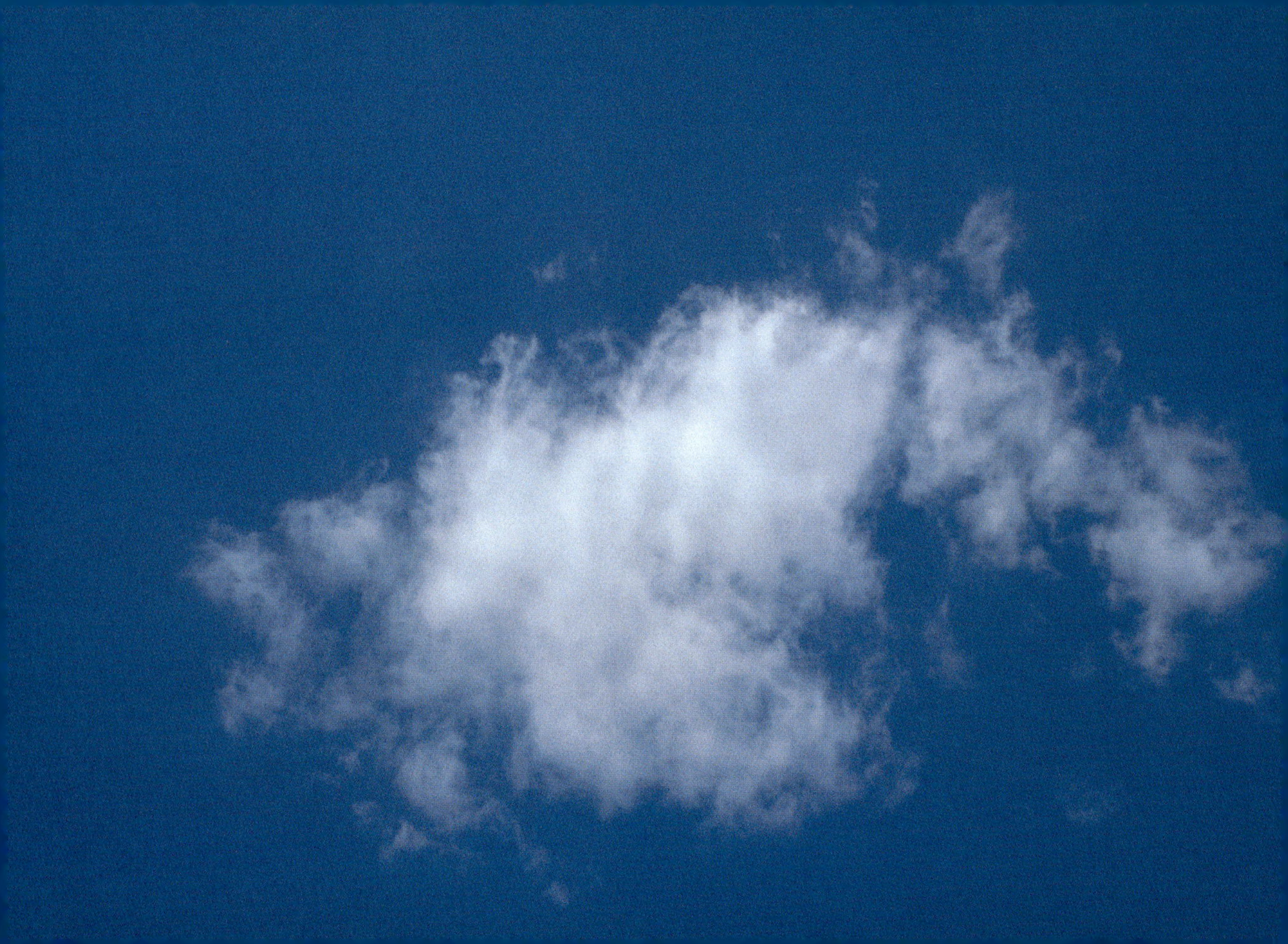

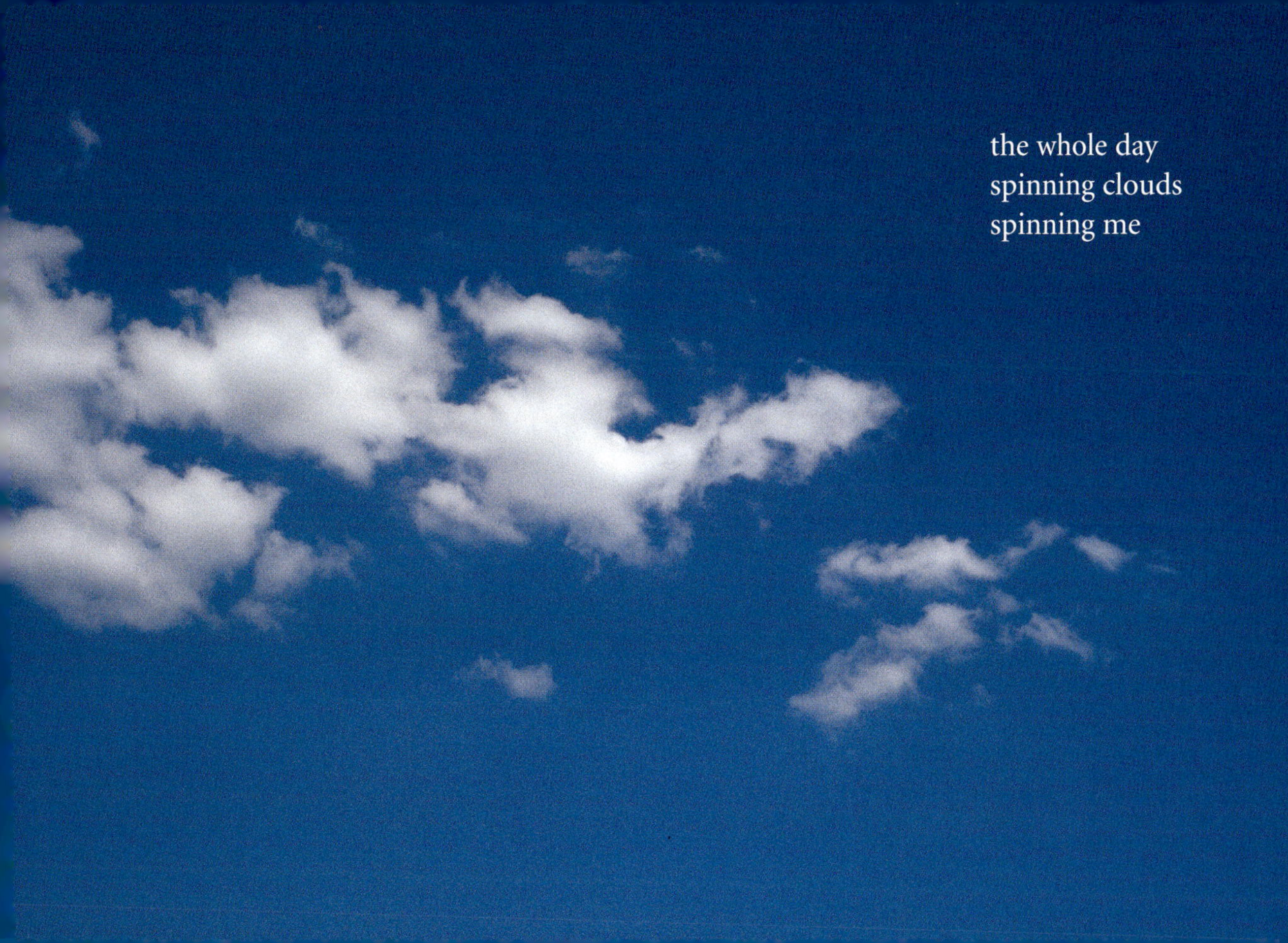
the whole day
spinning clouds
spinning me

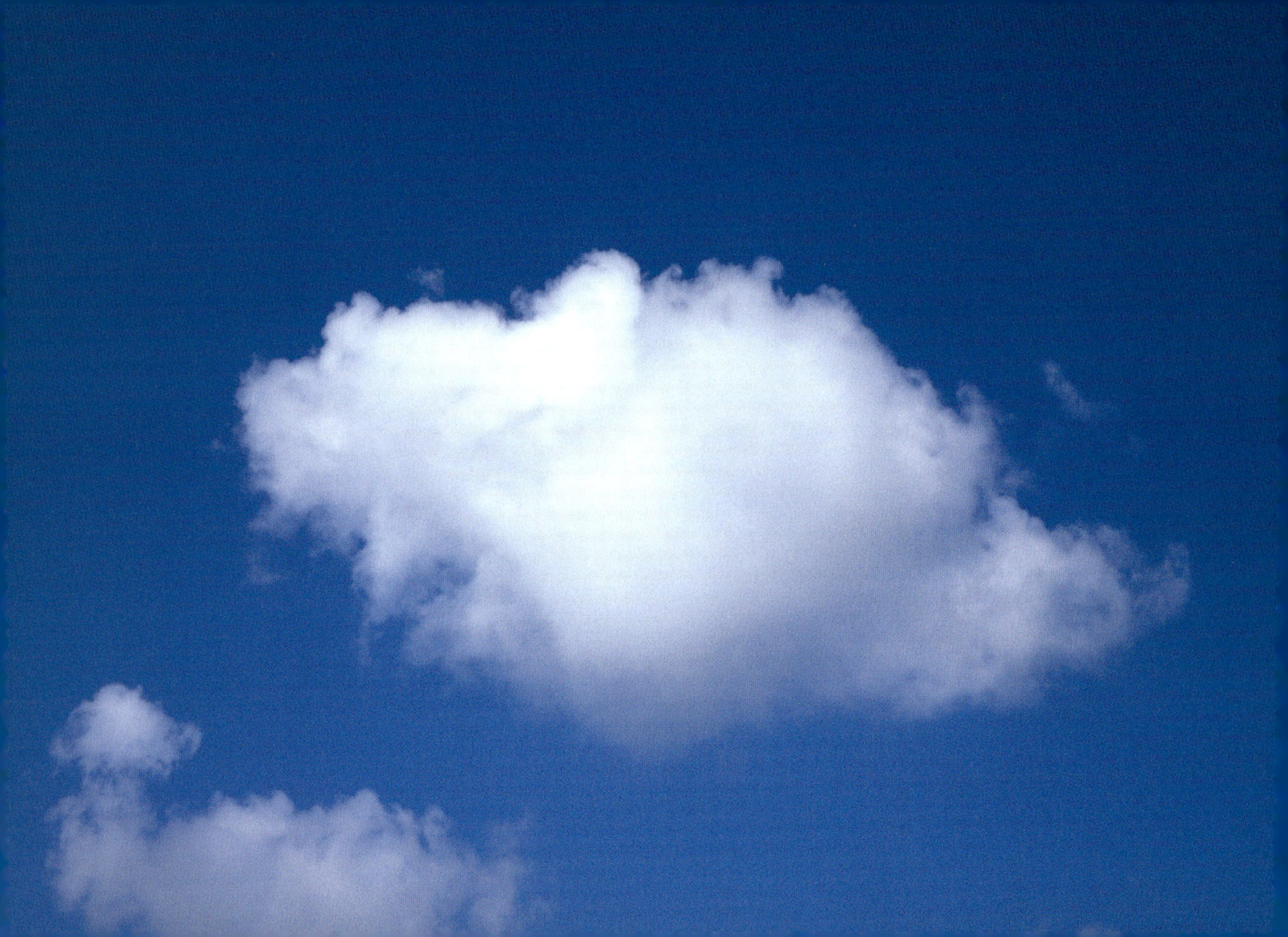

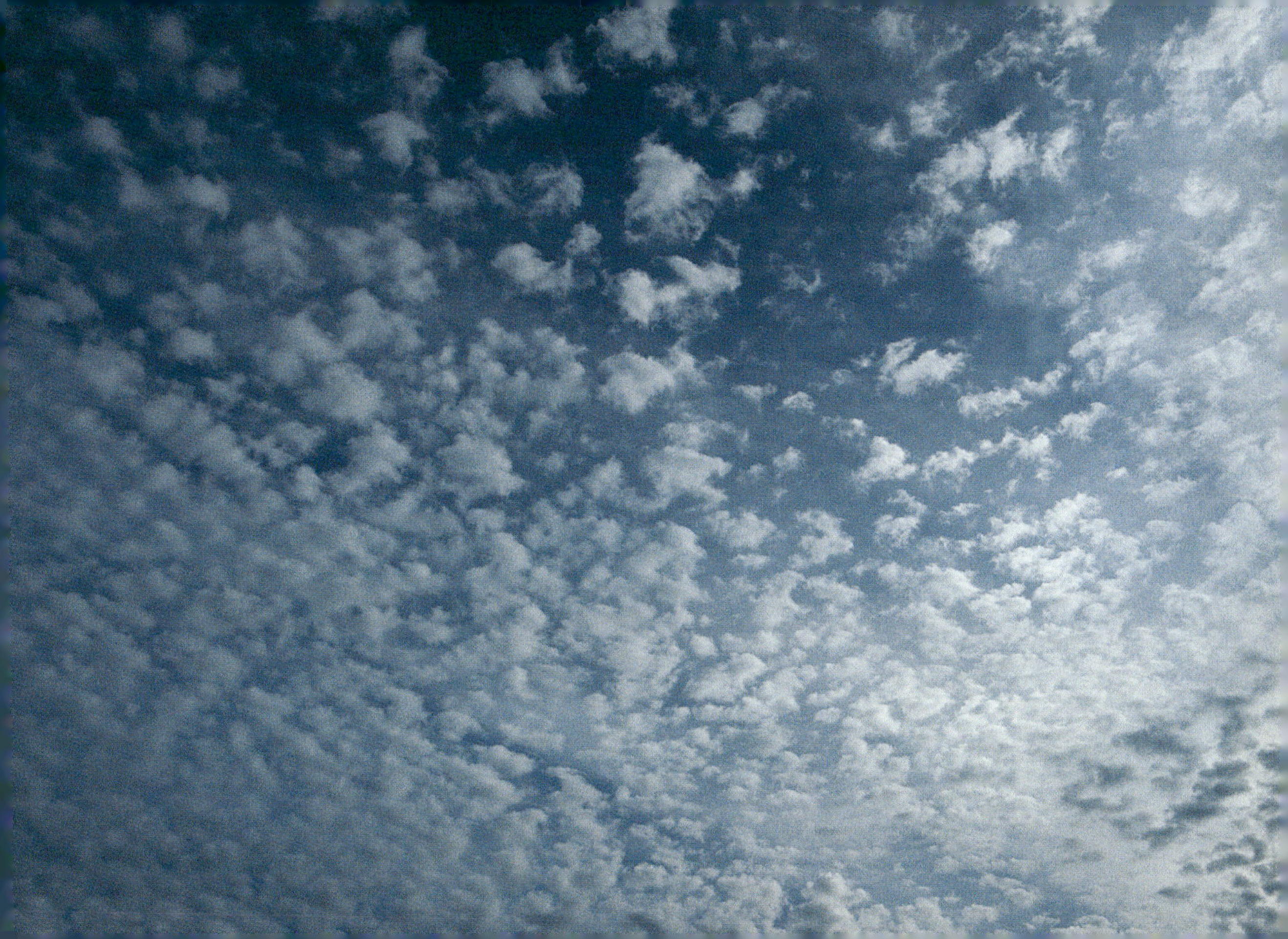

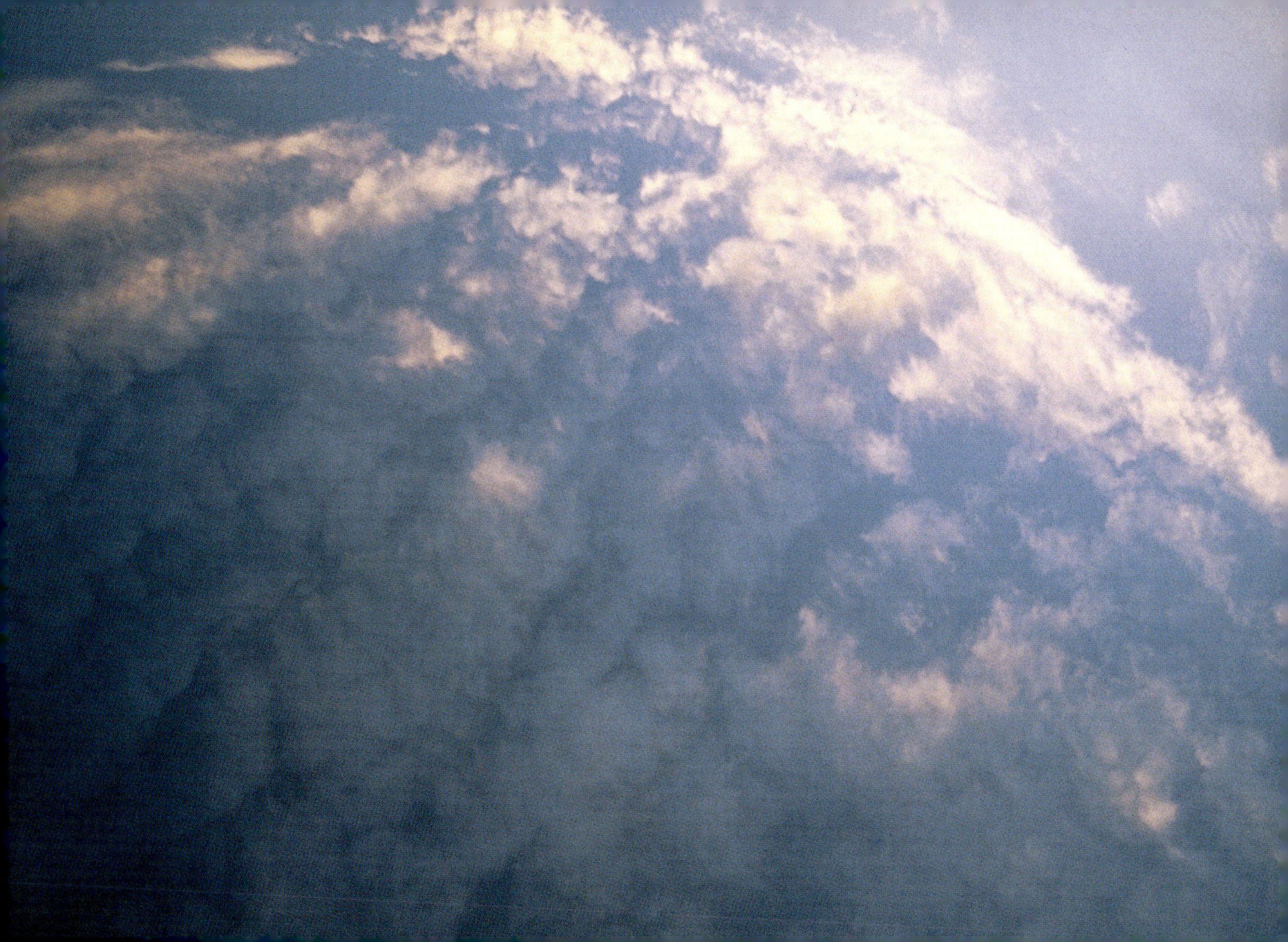

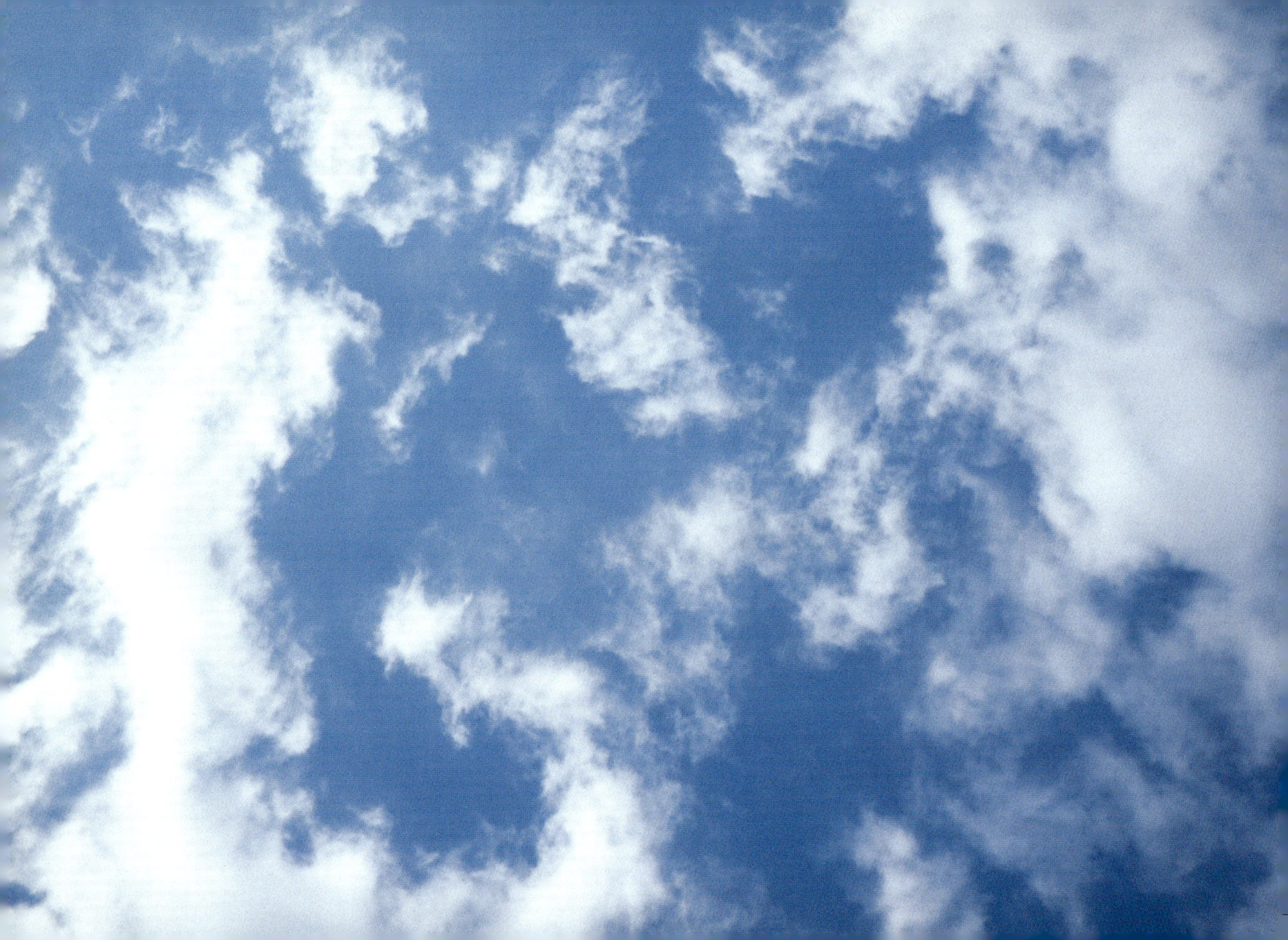

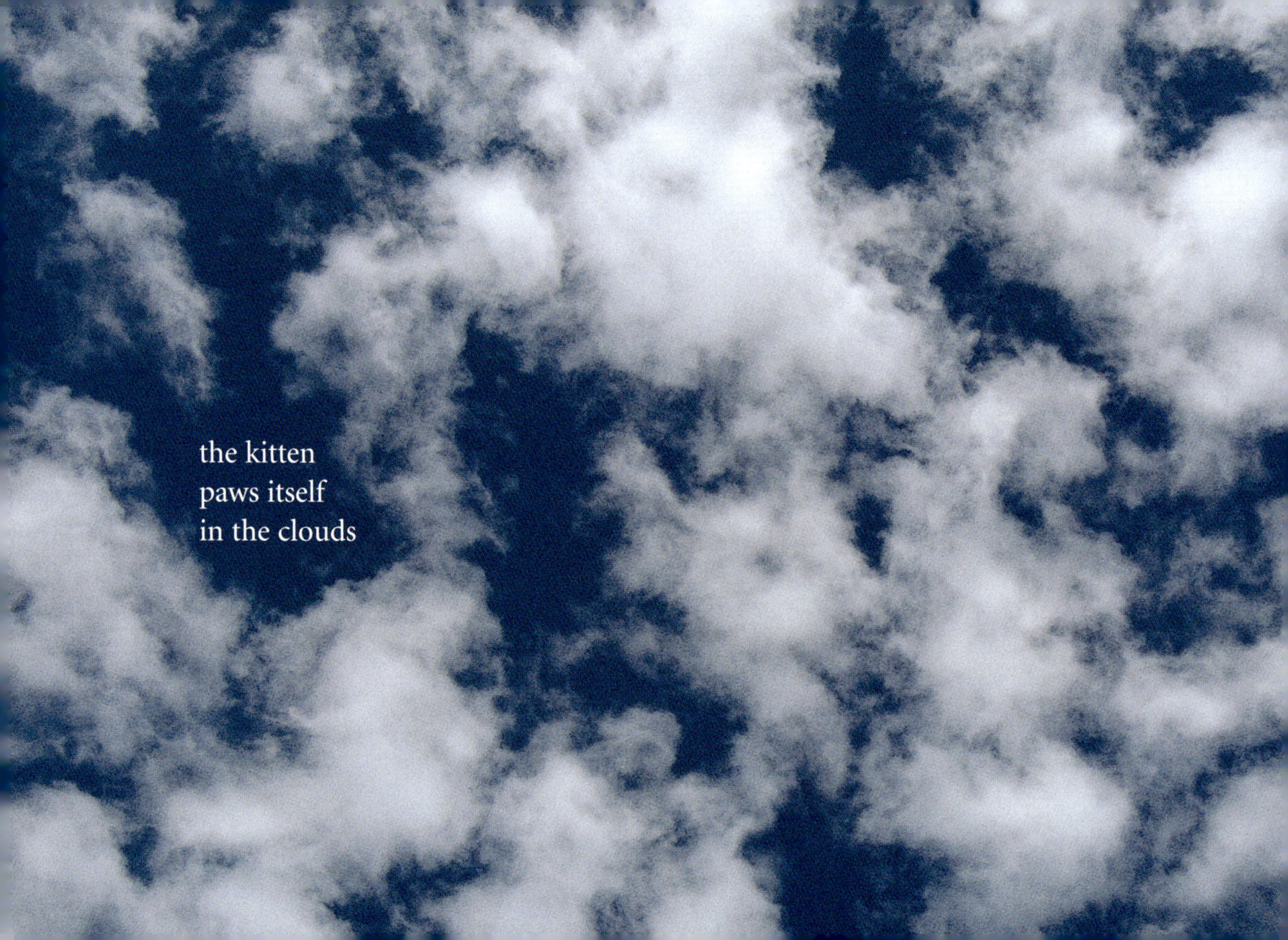
the kitten
paws itself
in the clouds

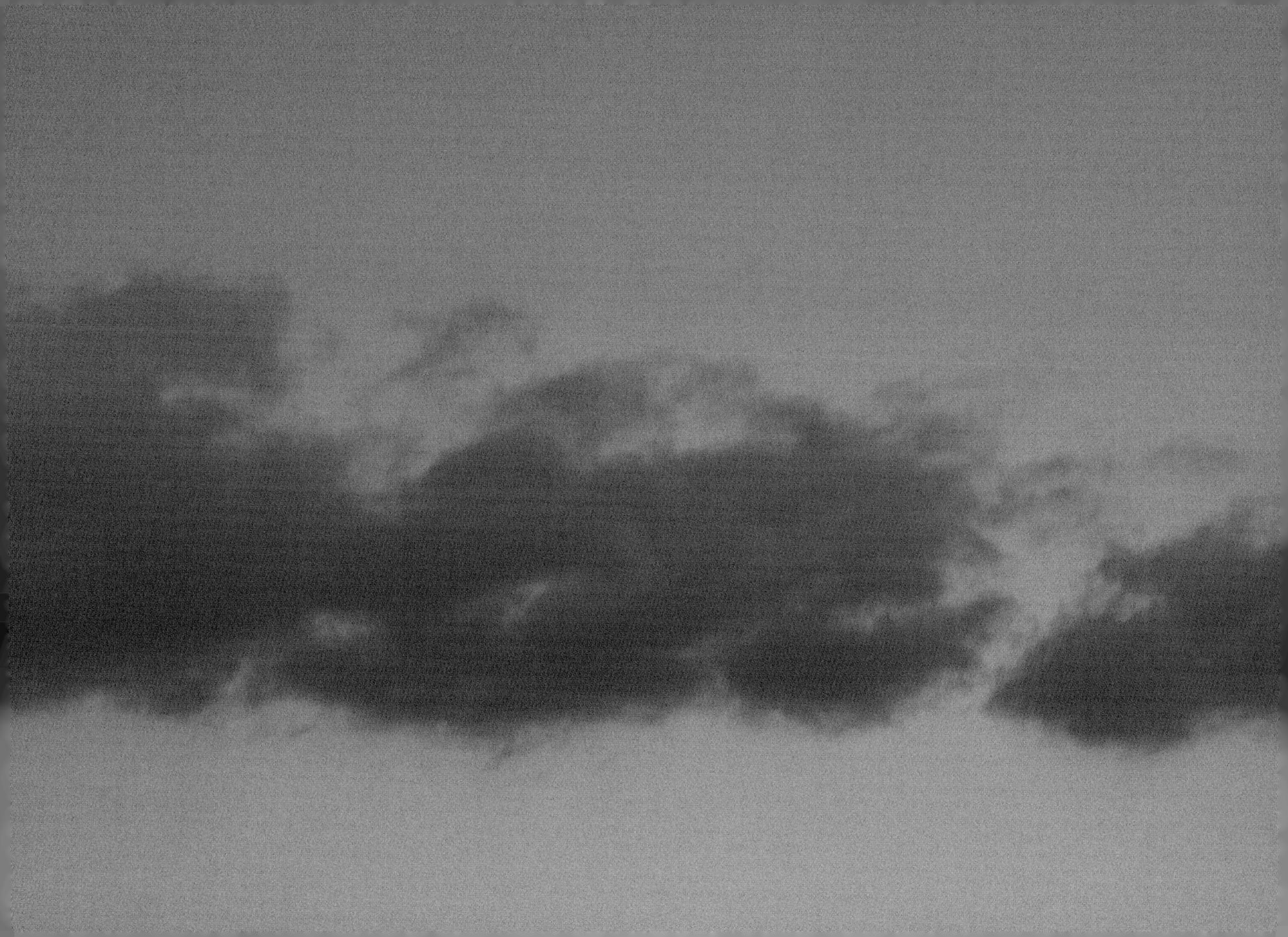

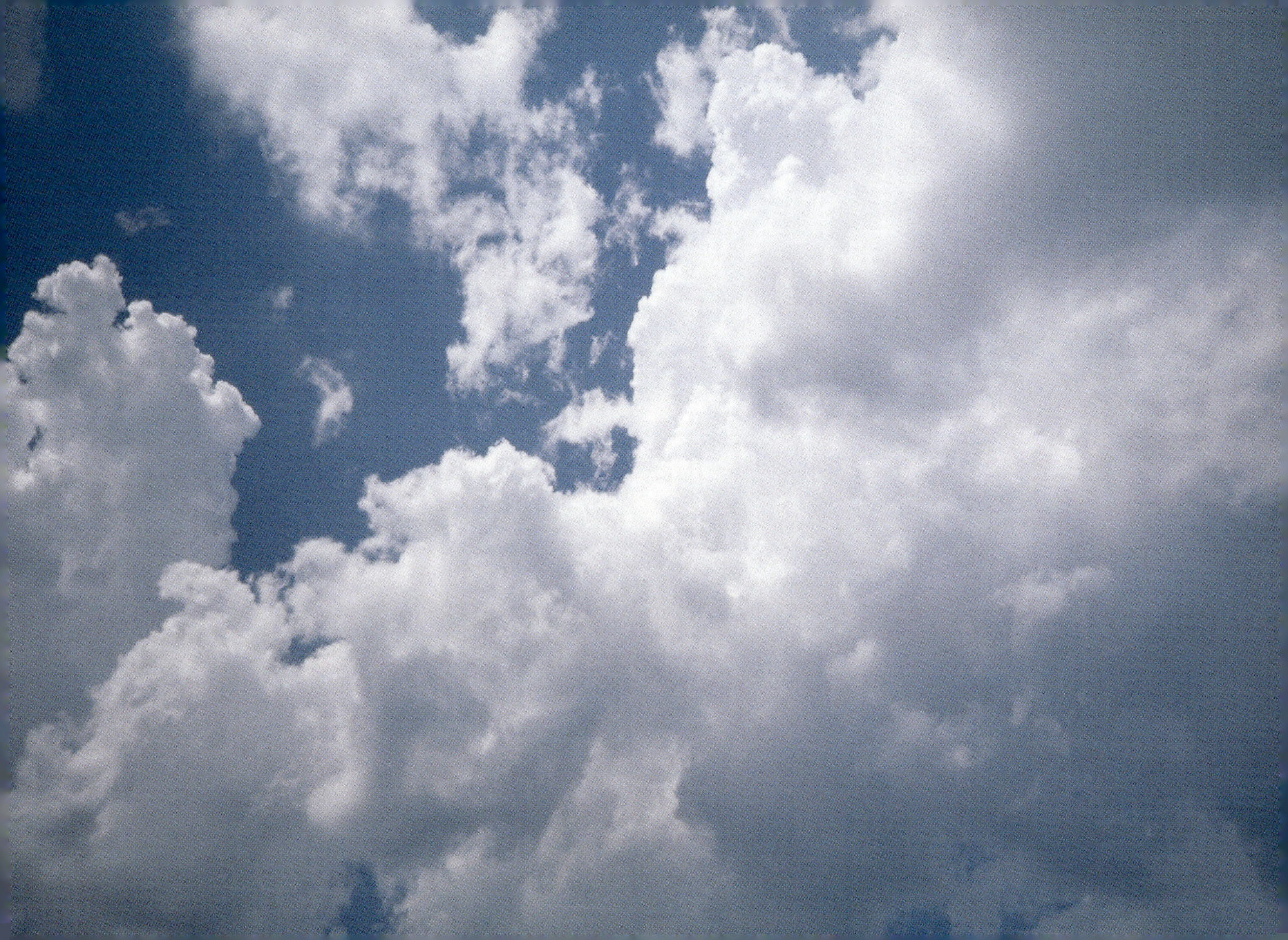

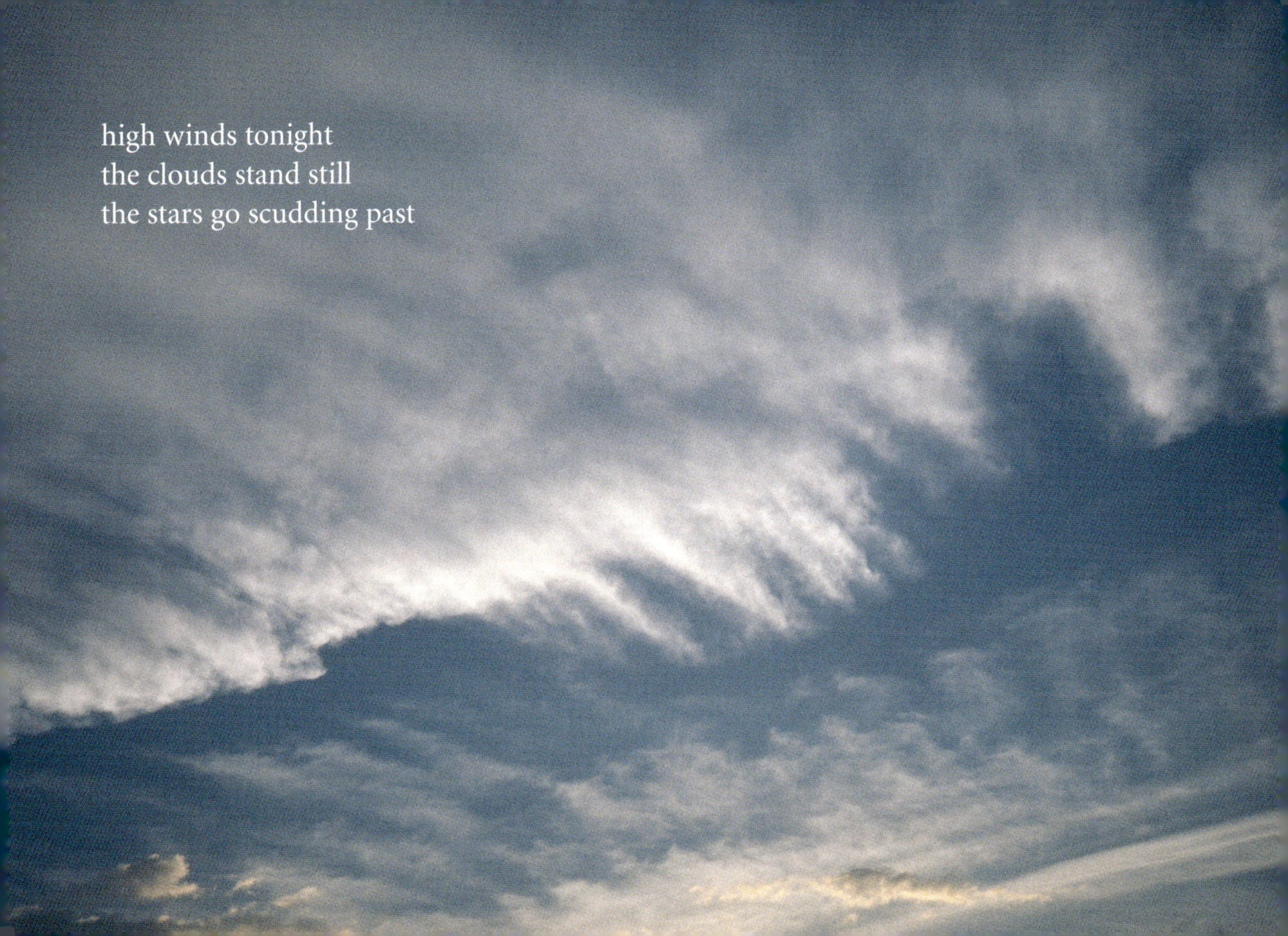
high winds tonight
the clouds stand still
the stars go scudding past

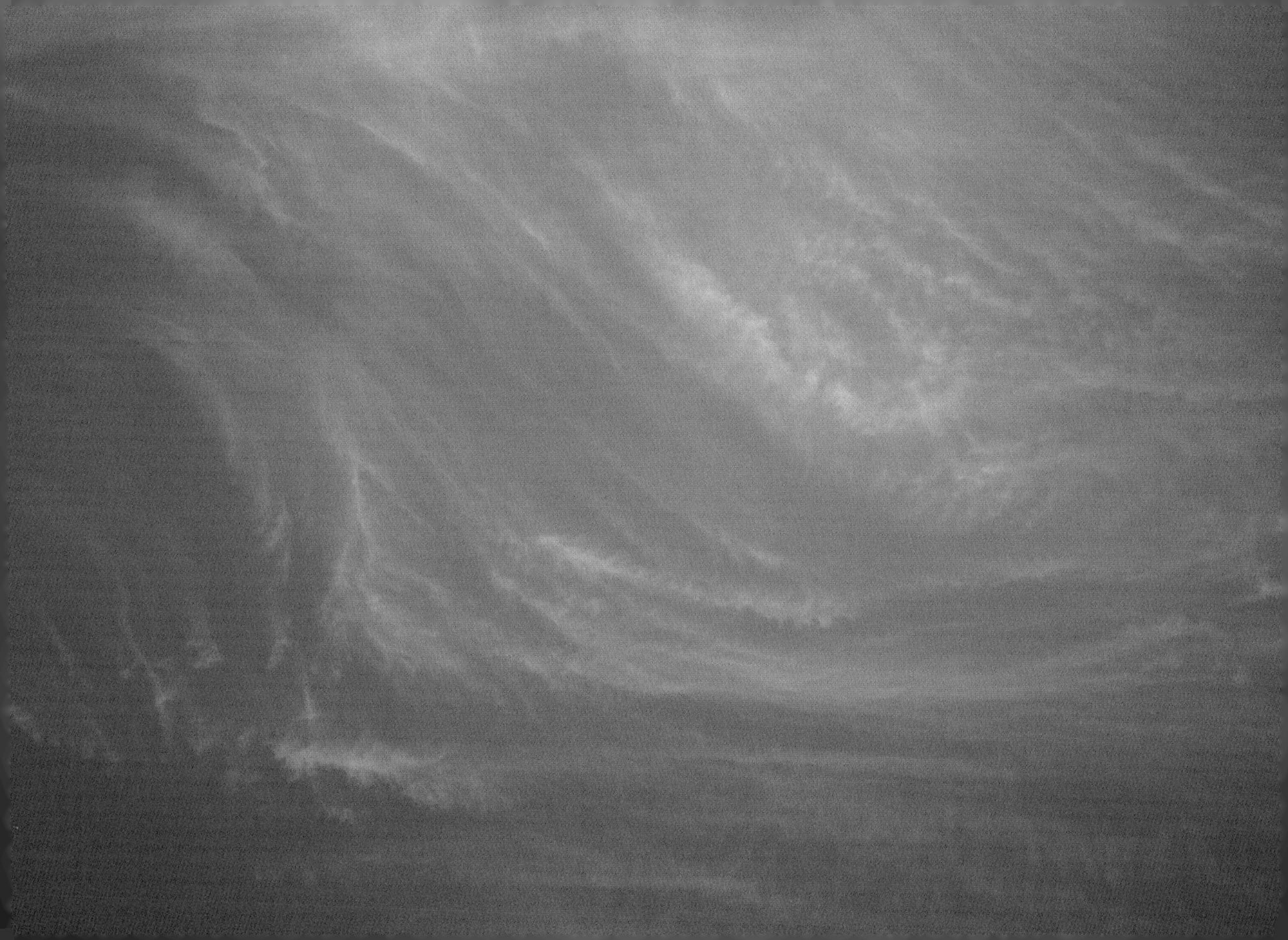

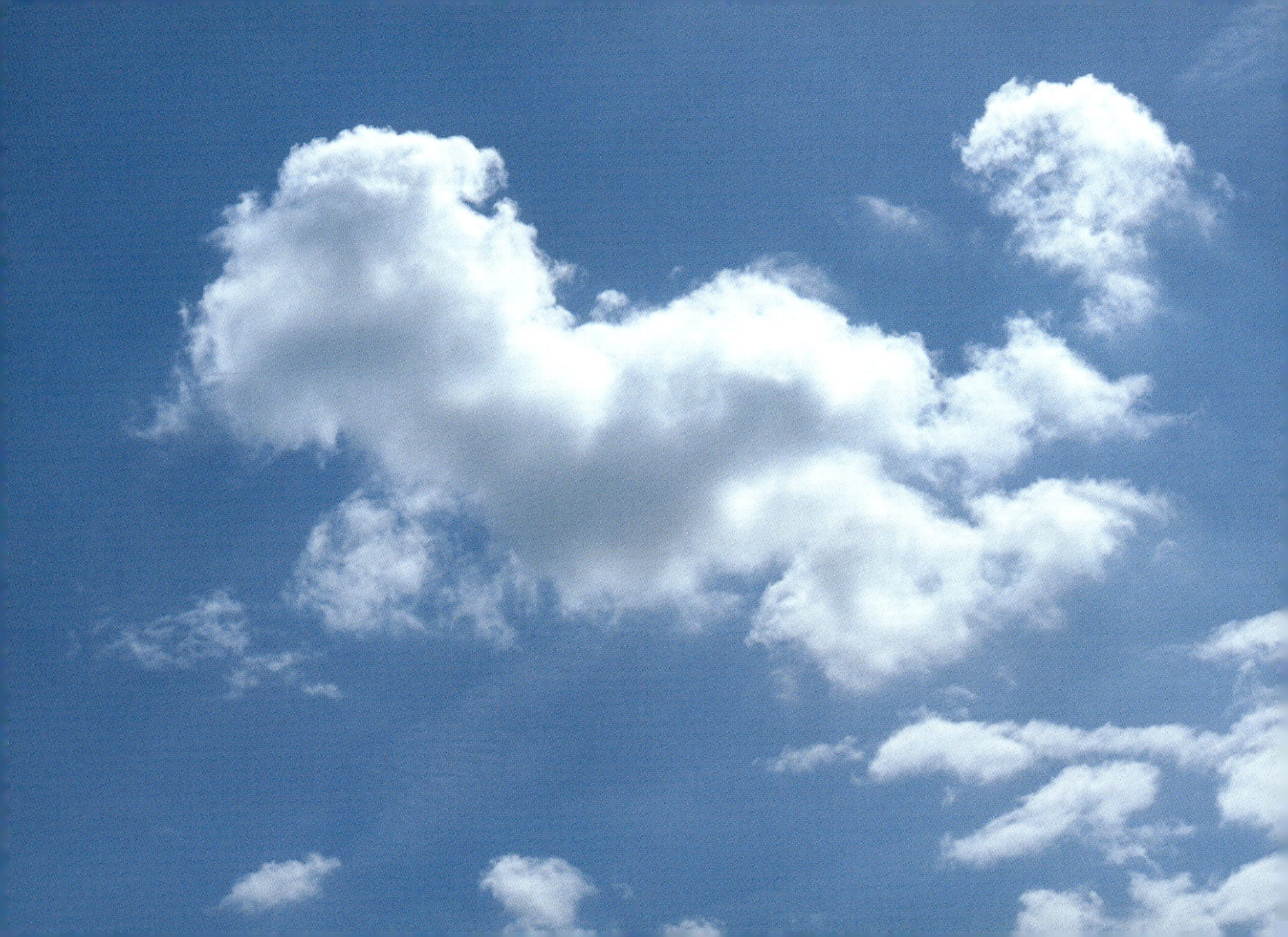

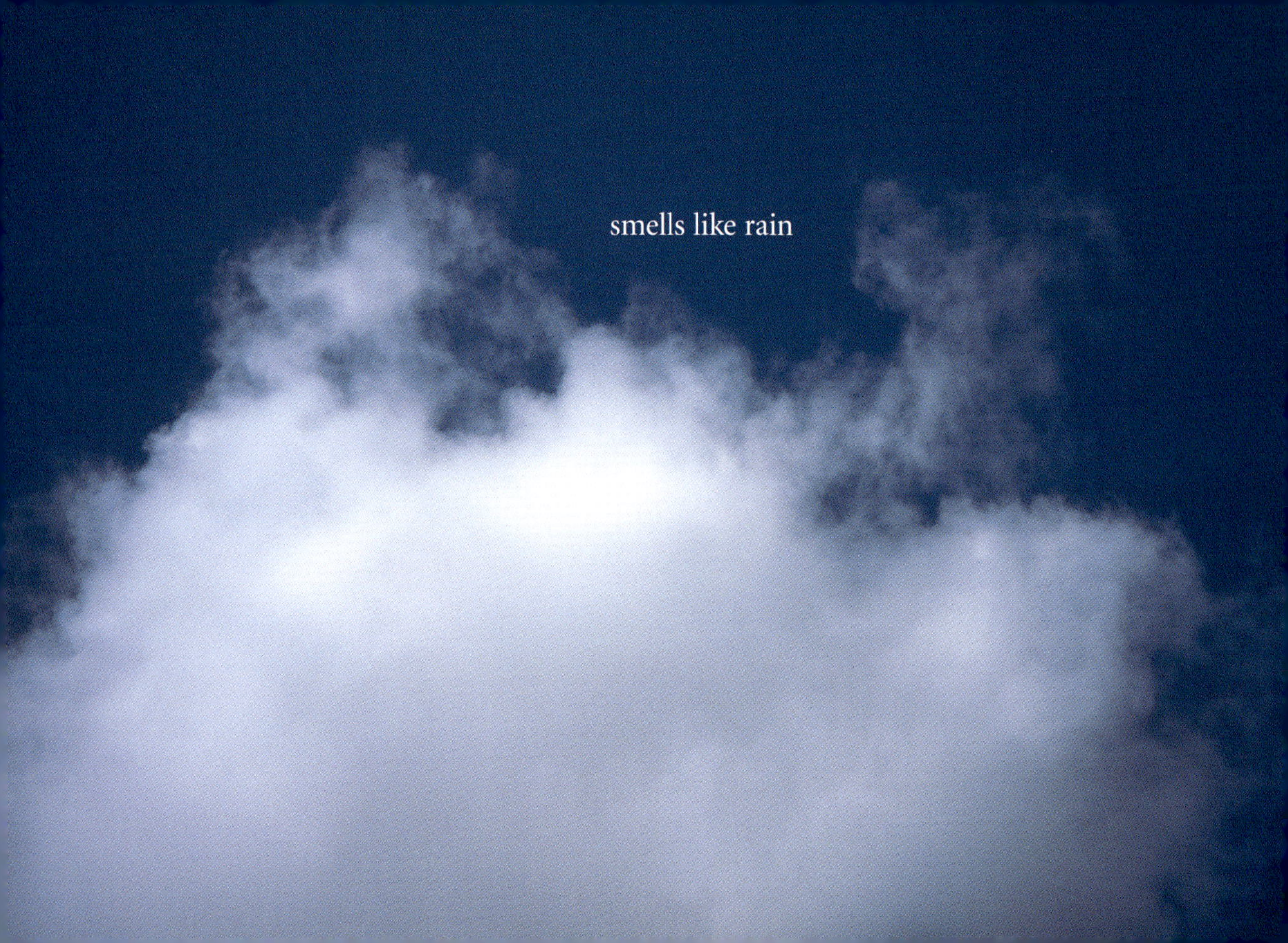

smells like rain

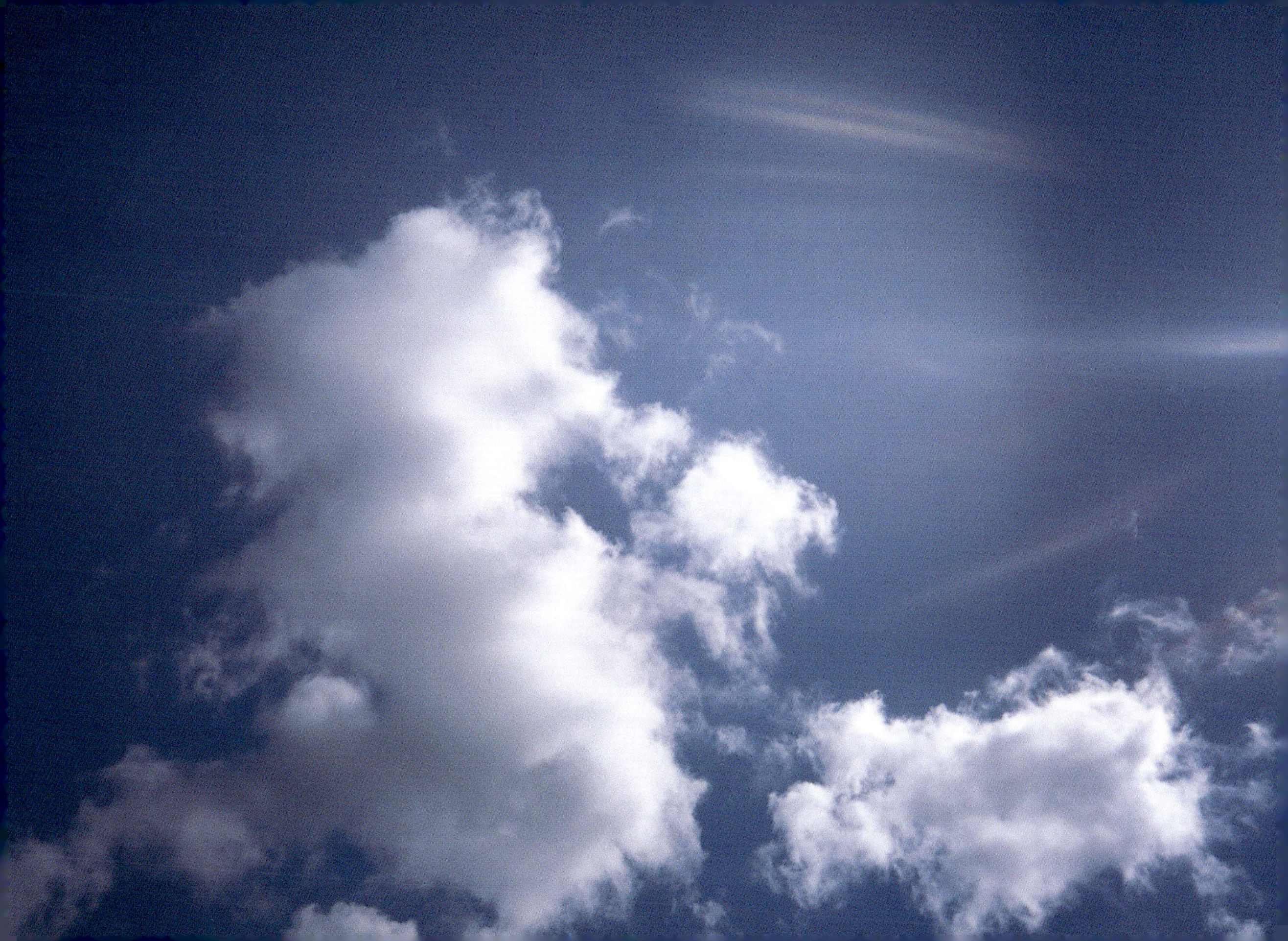

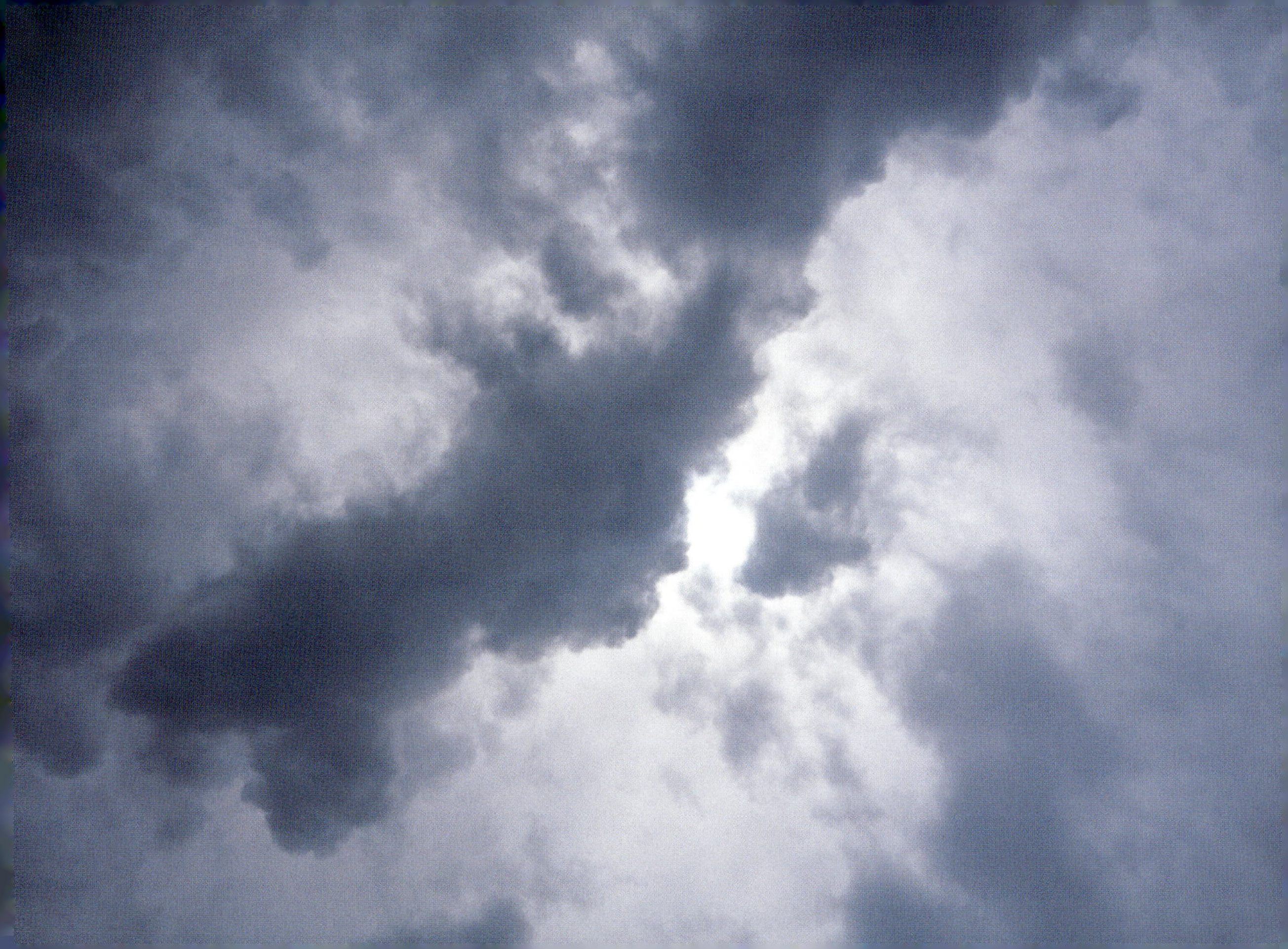

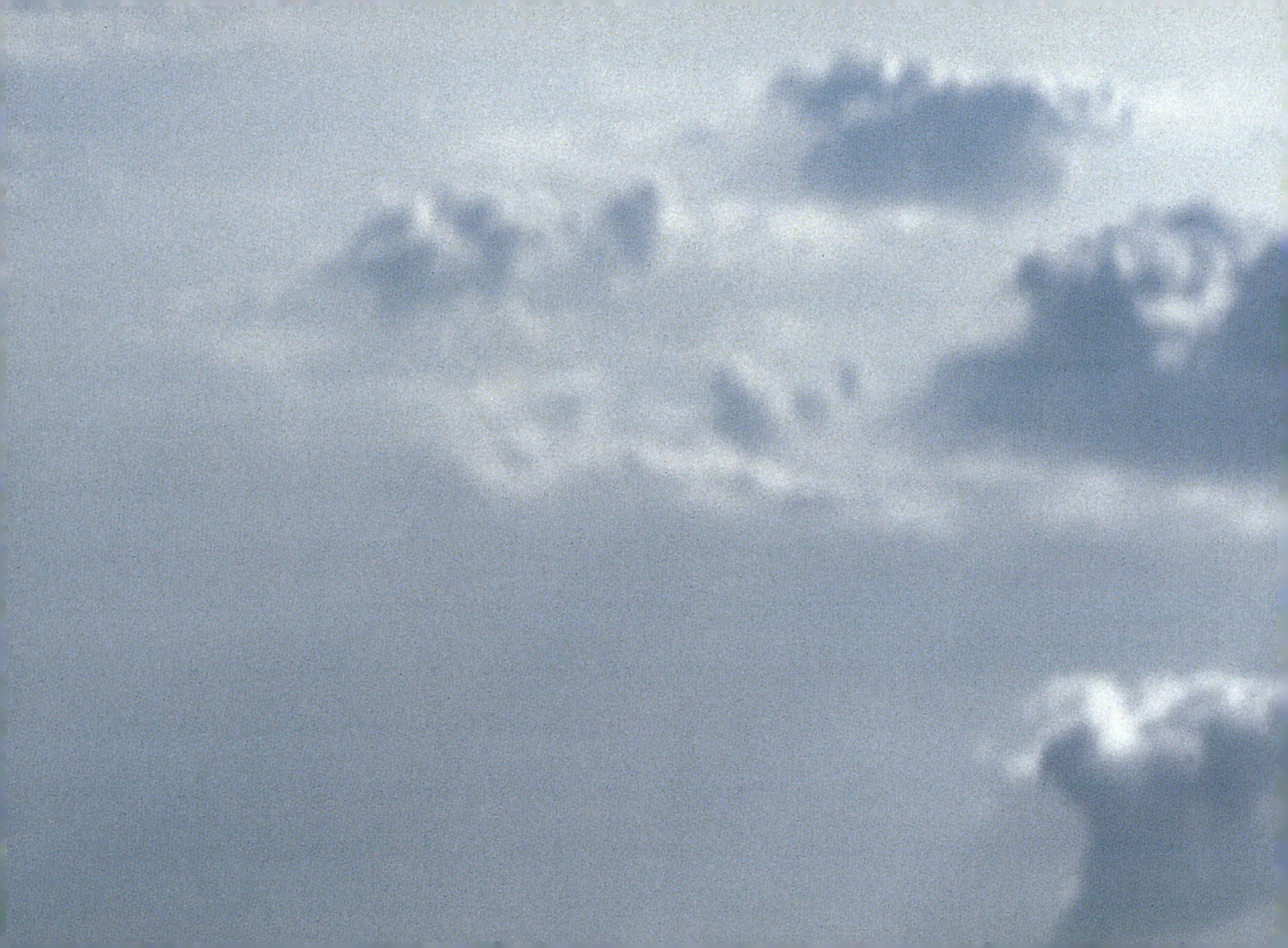

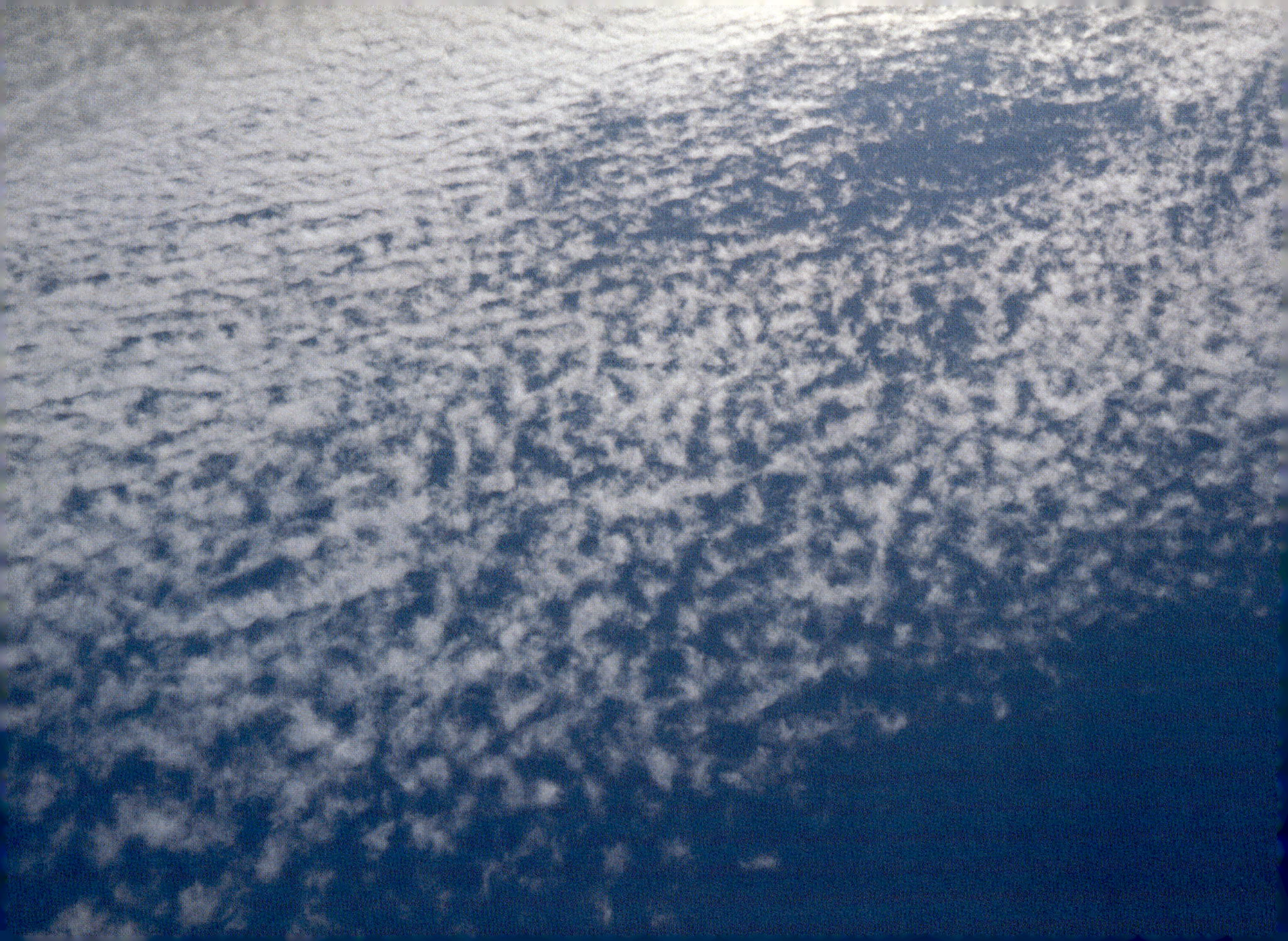

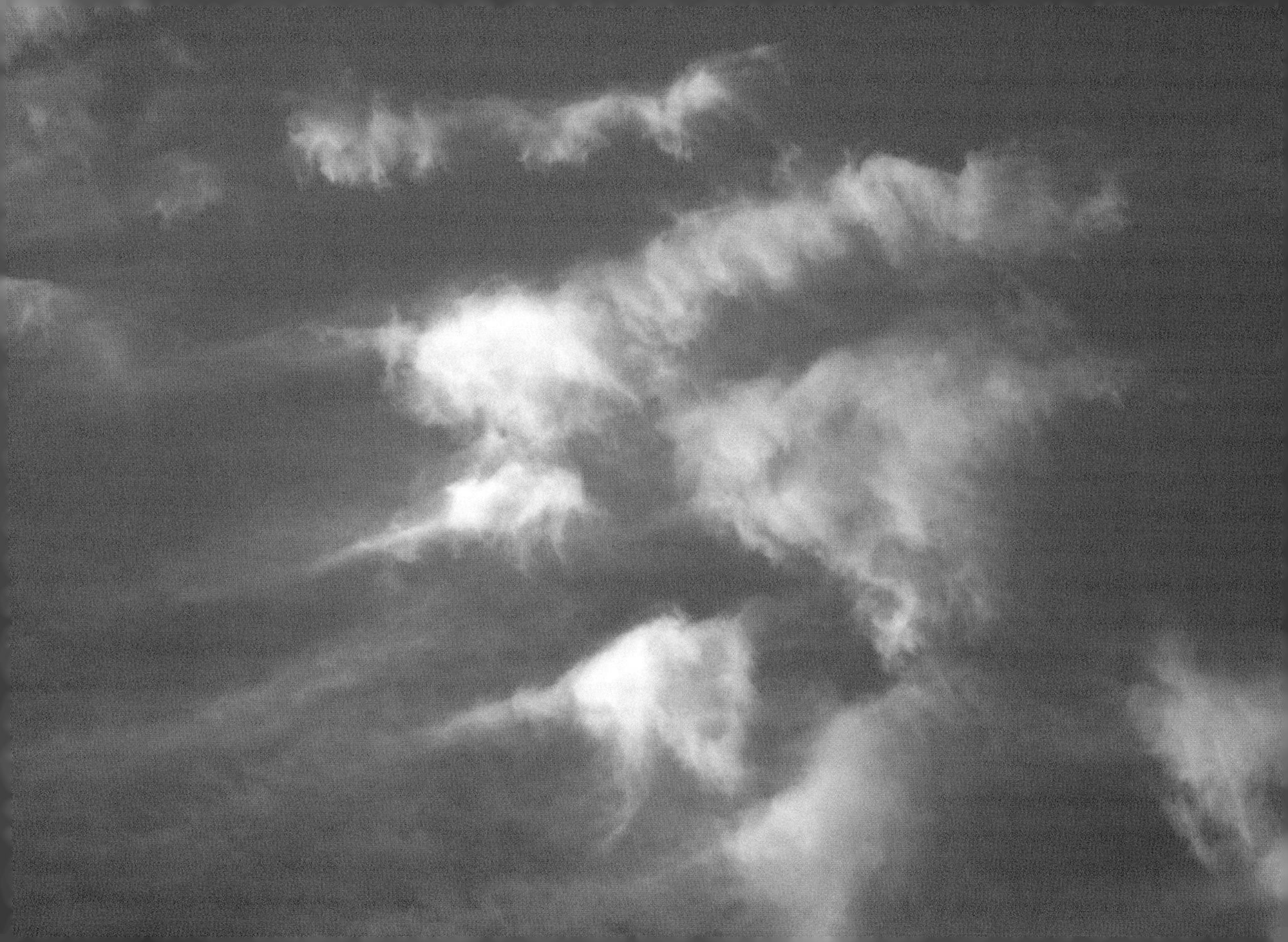

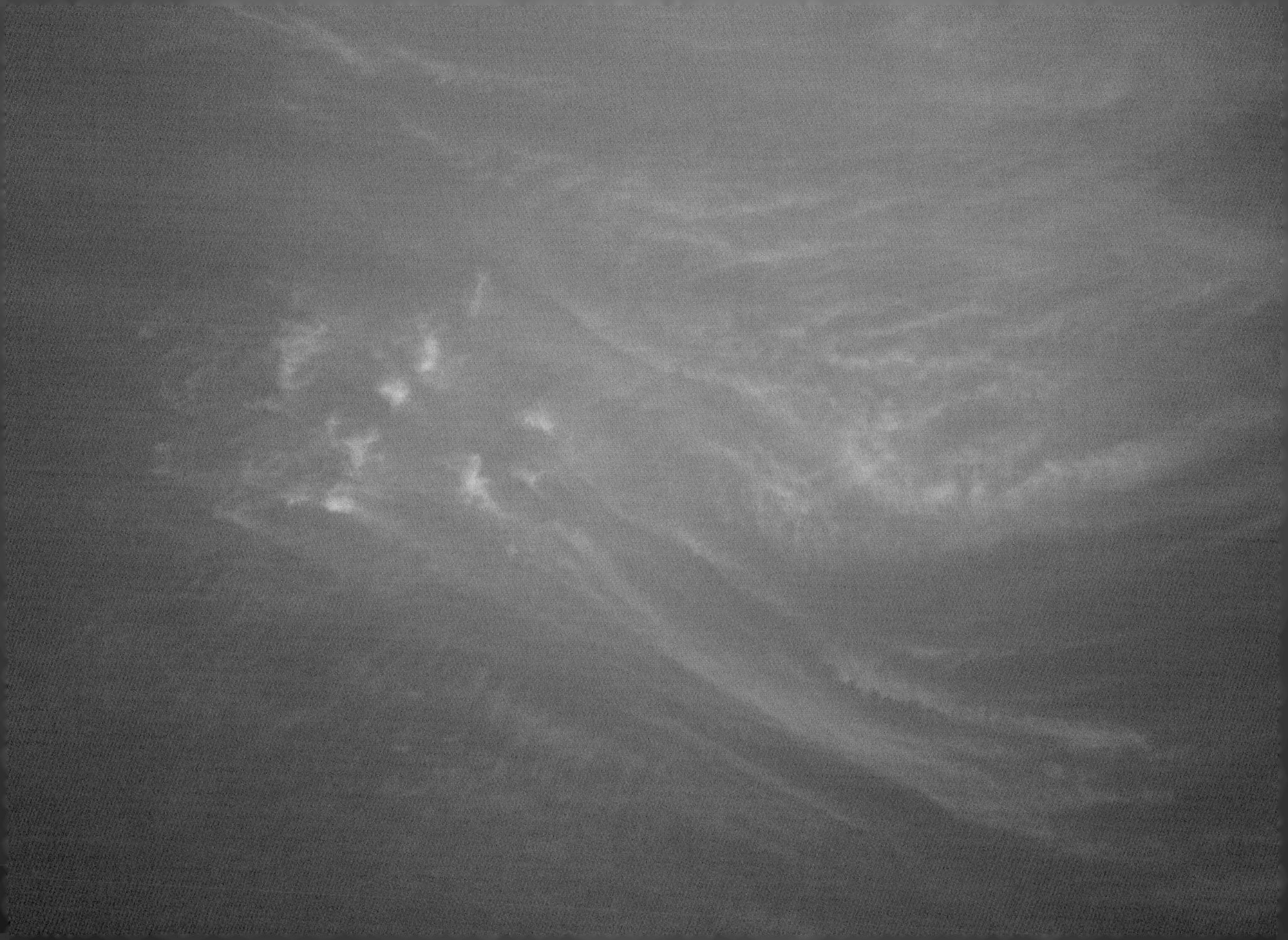

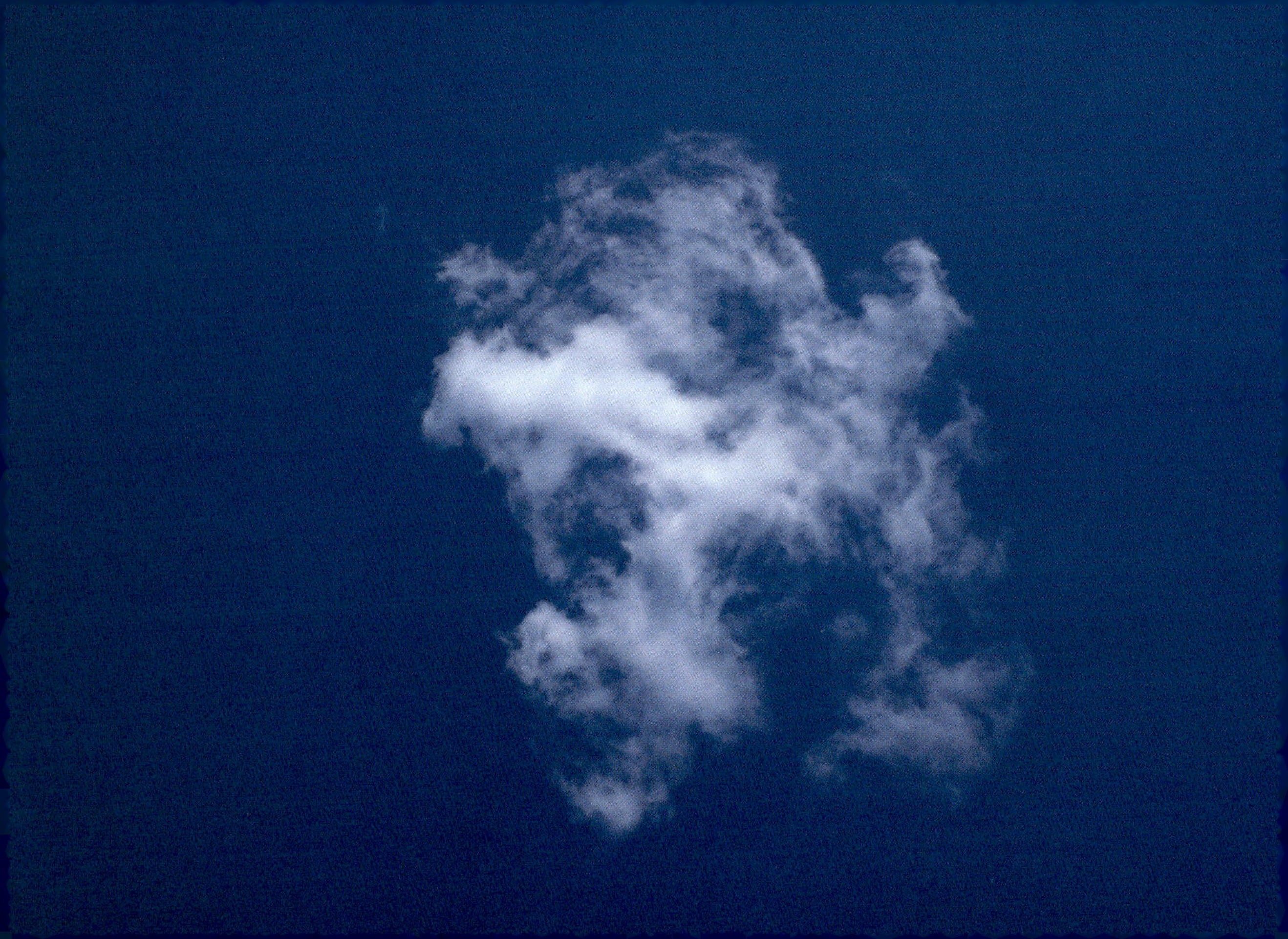

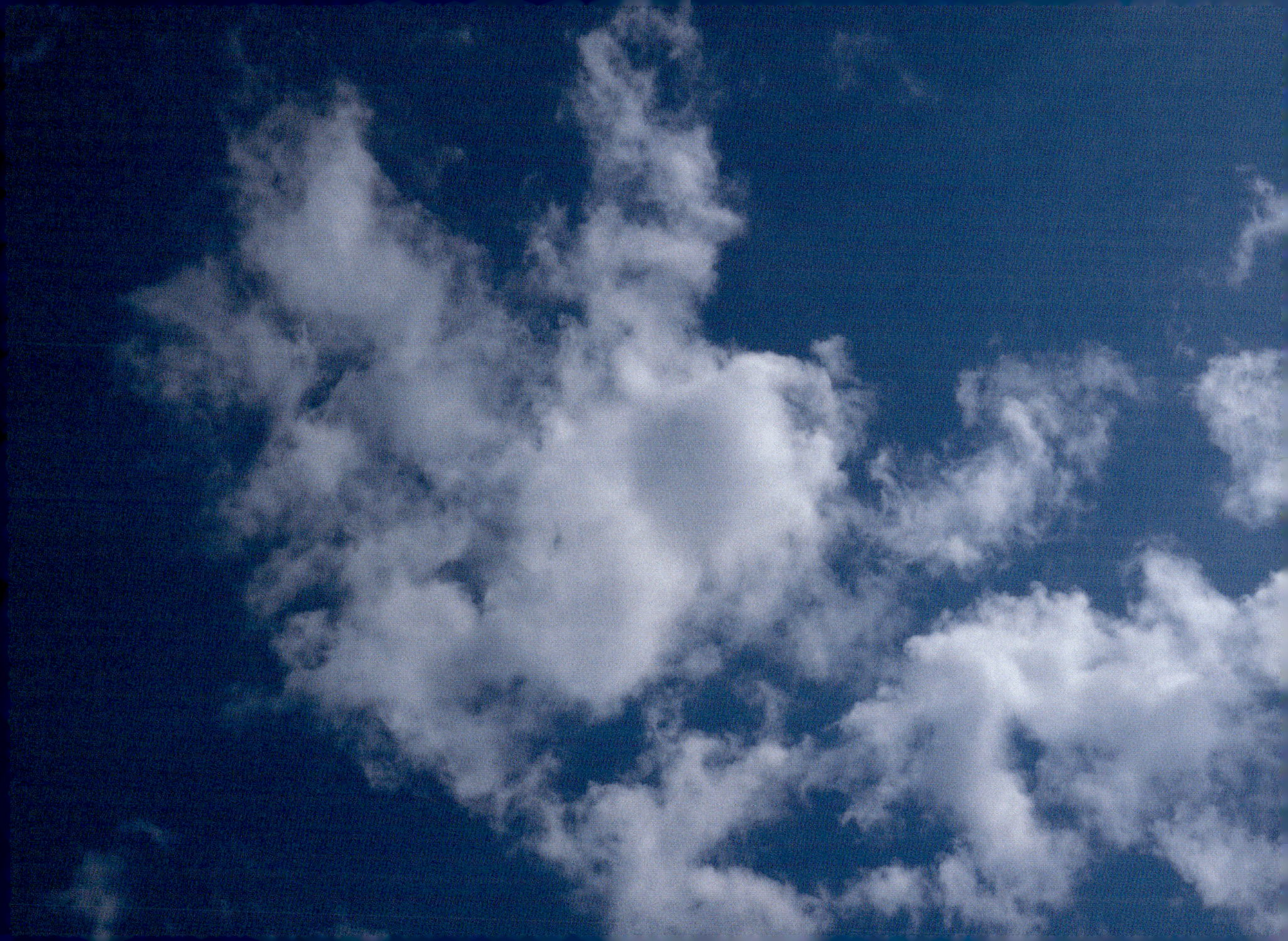

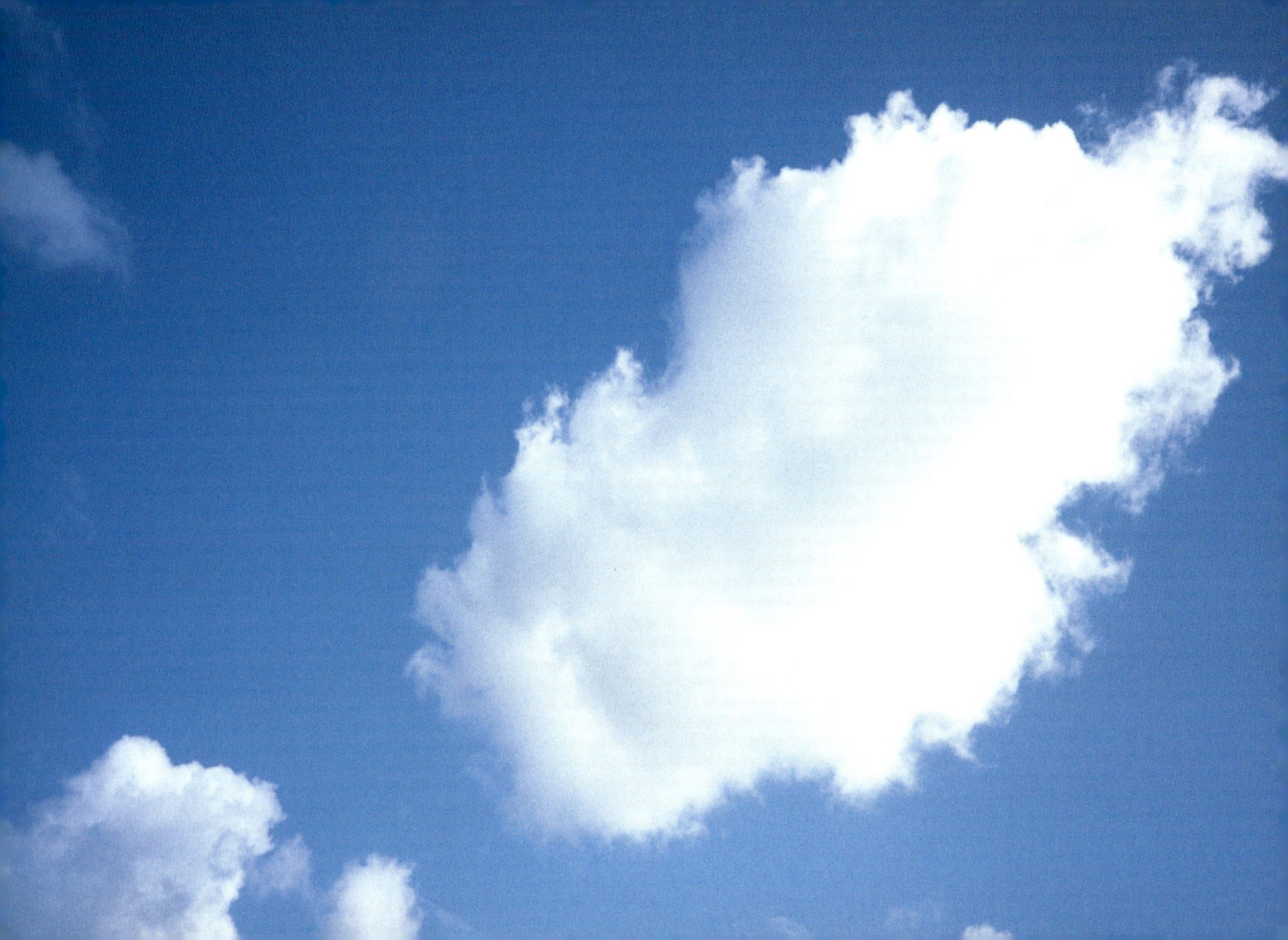

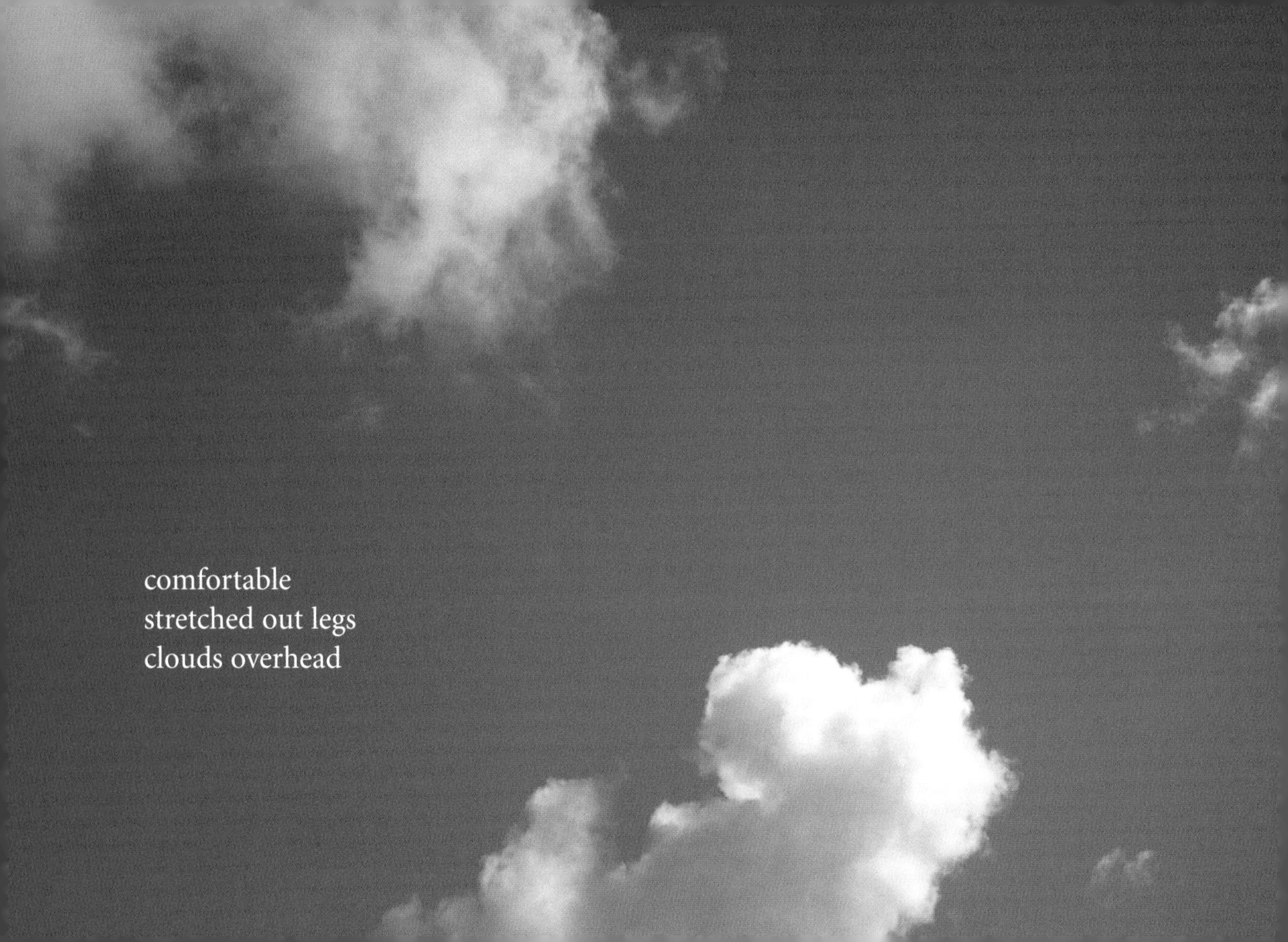

comfortable
stretched out legs
clouds overhead

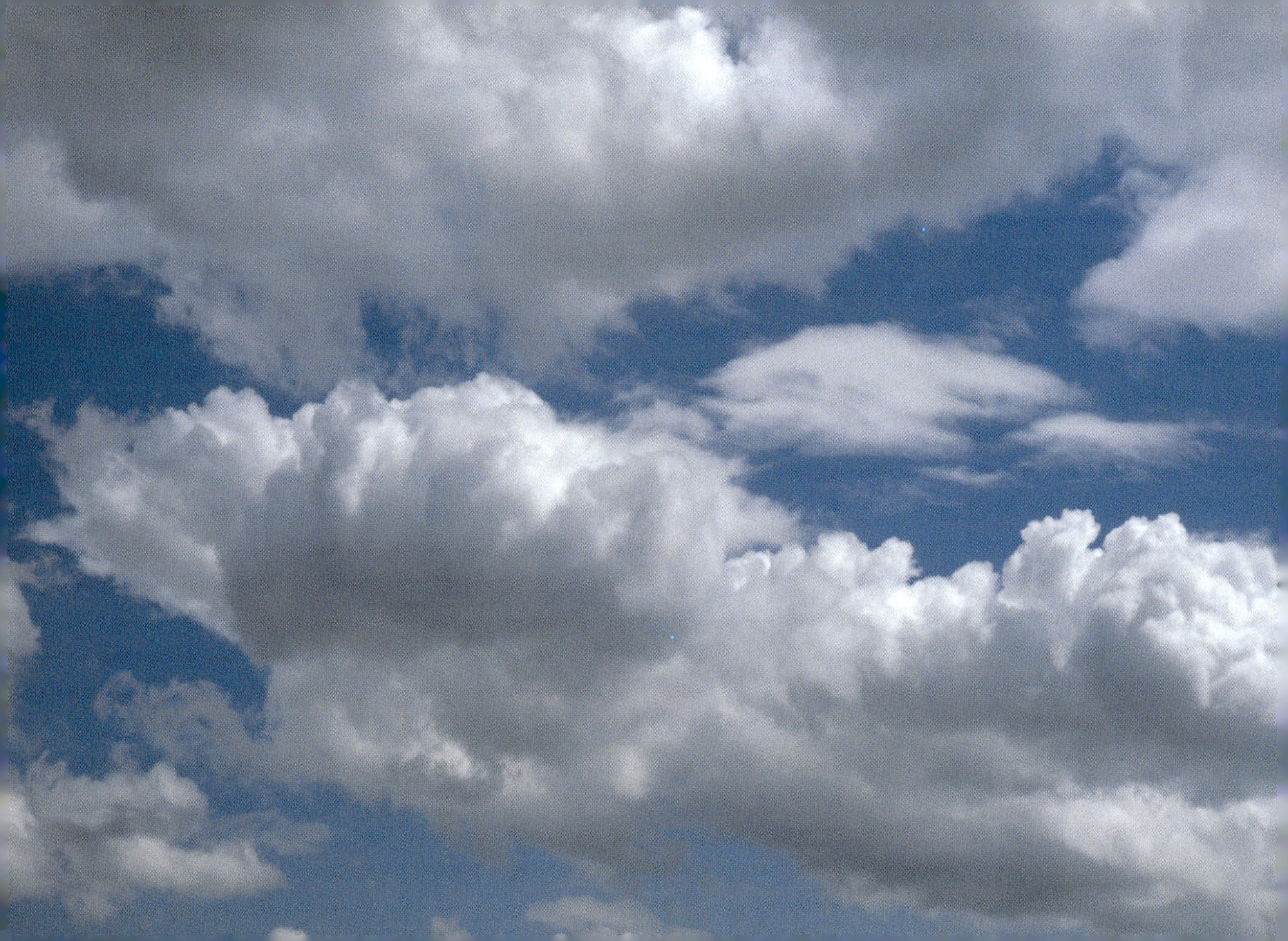

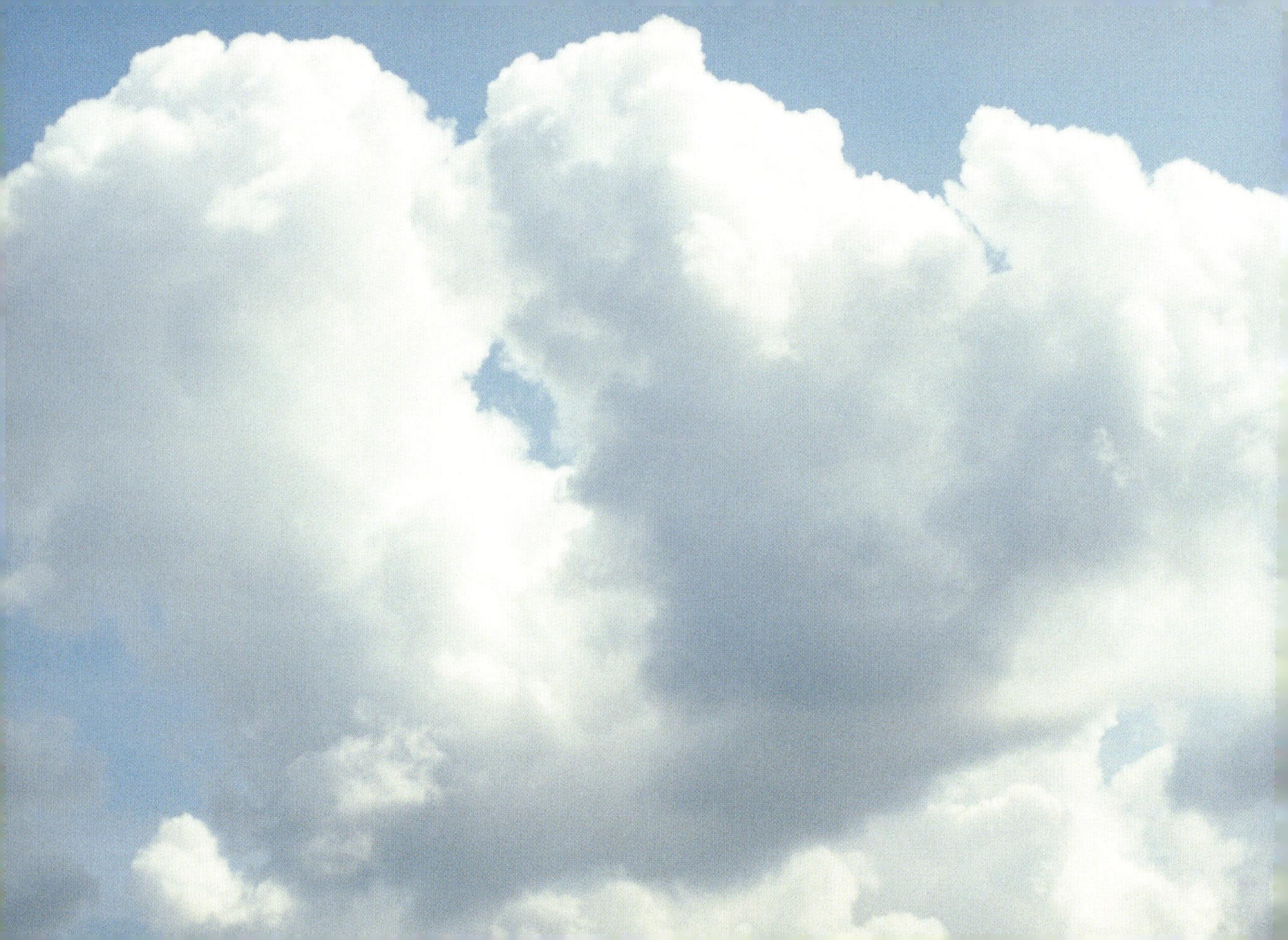

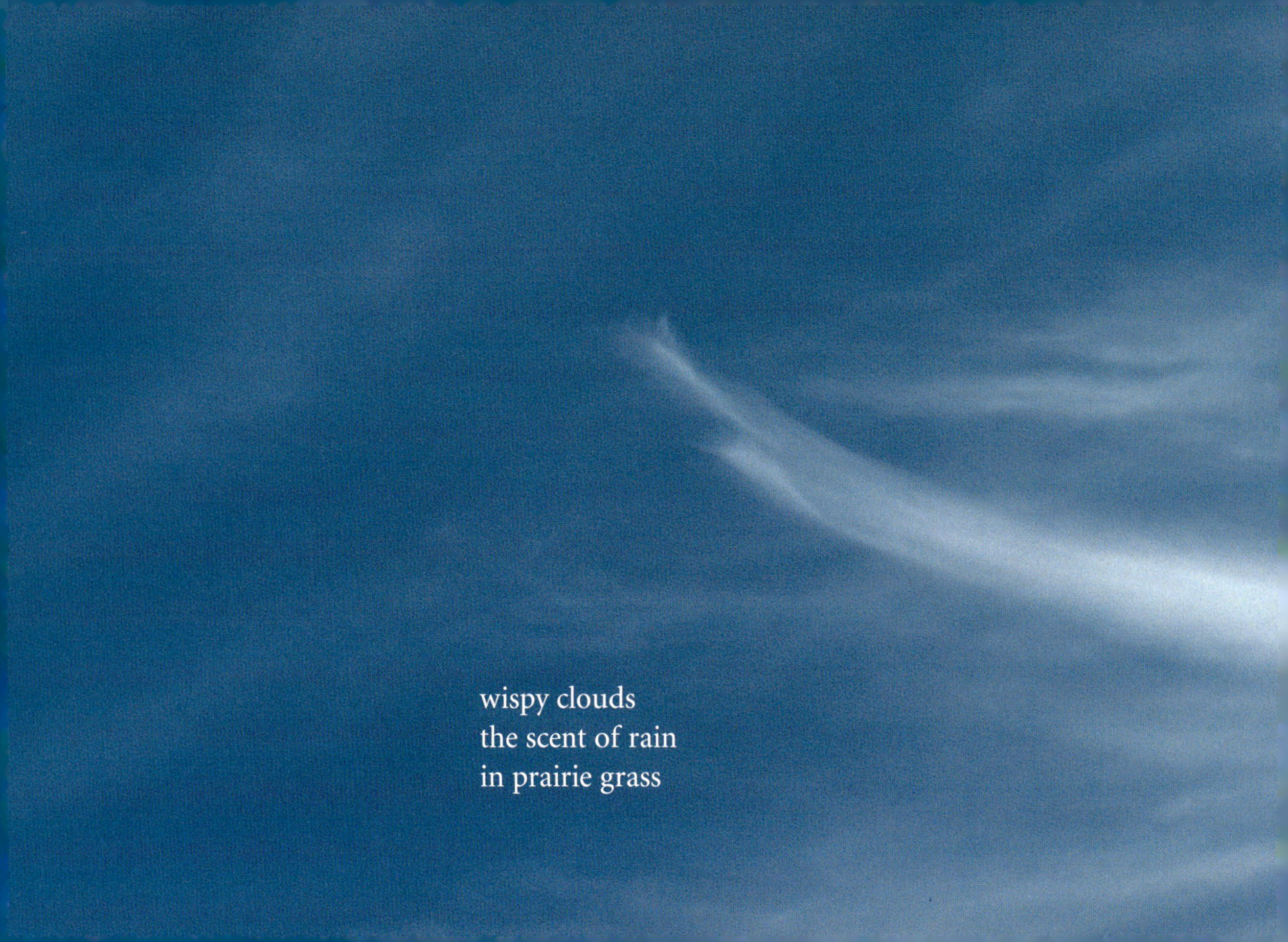

wispy clouds
the scent of rain
in prairie grass

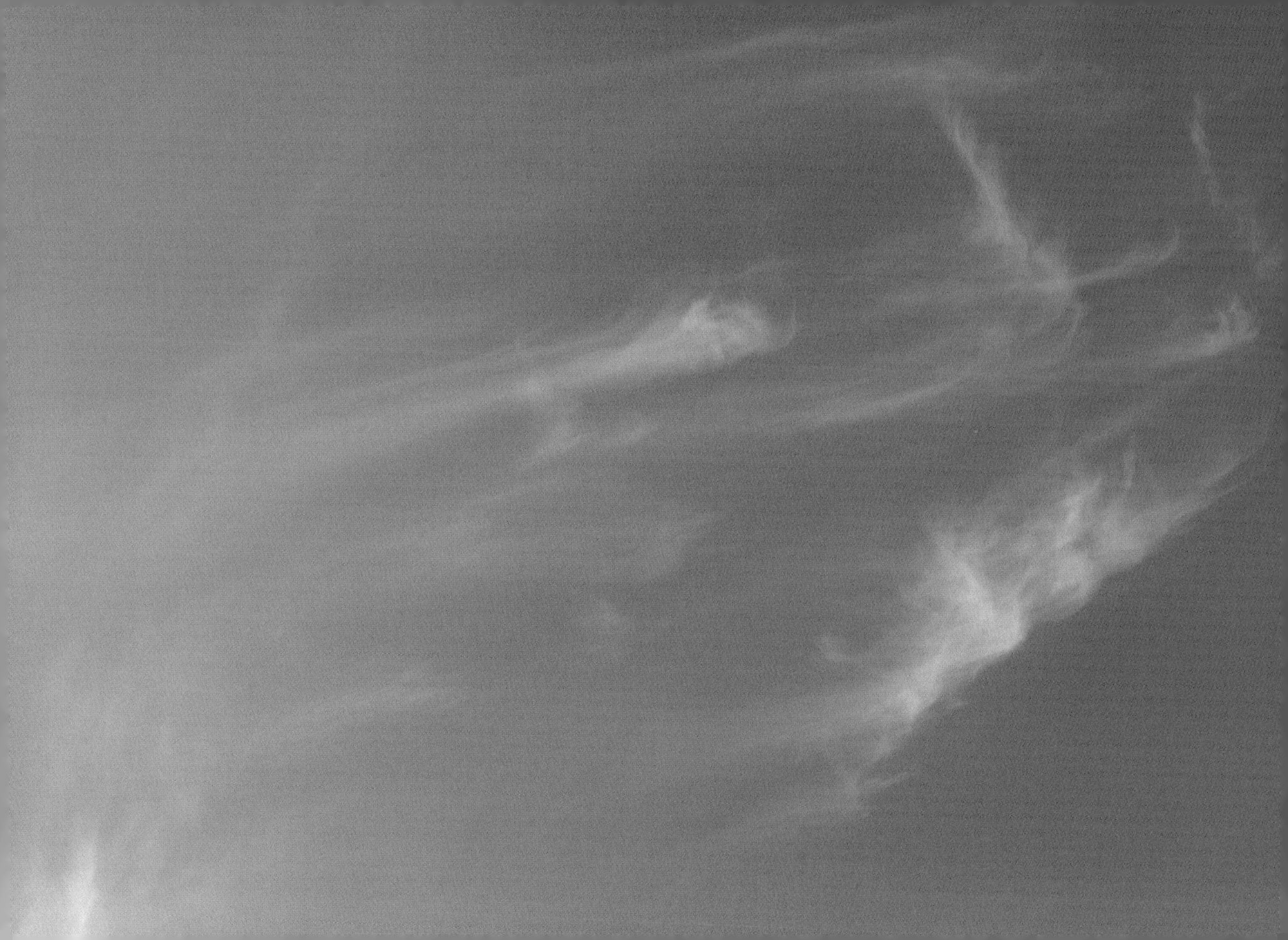

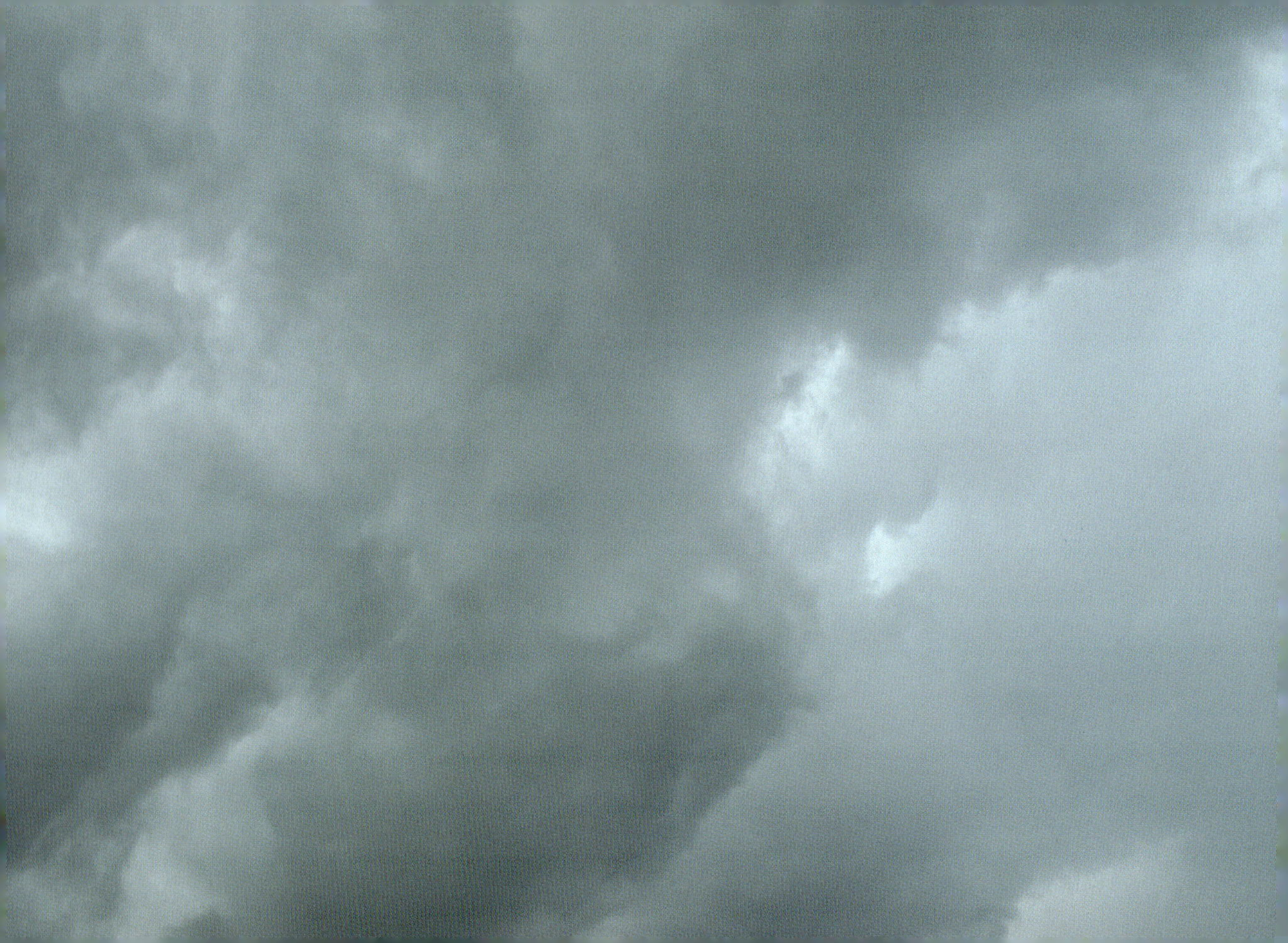

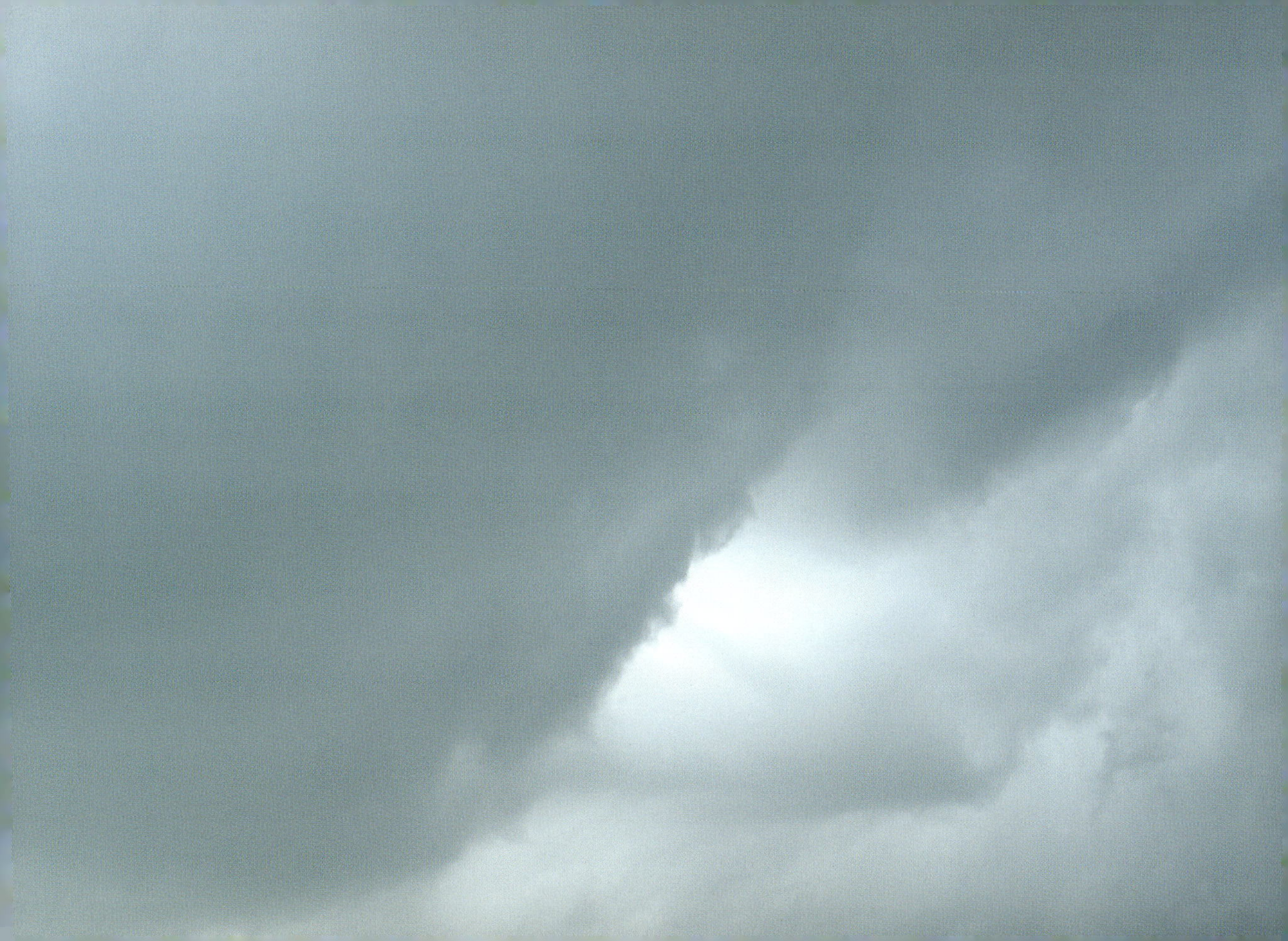

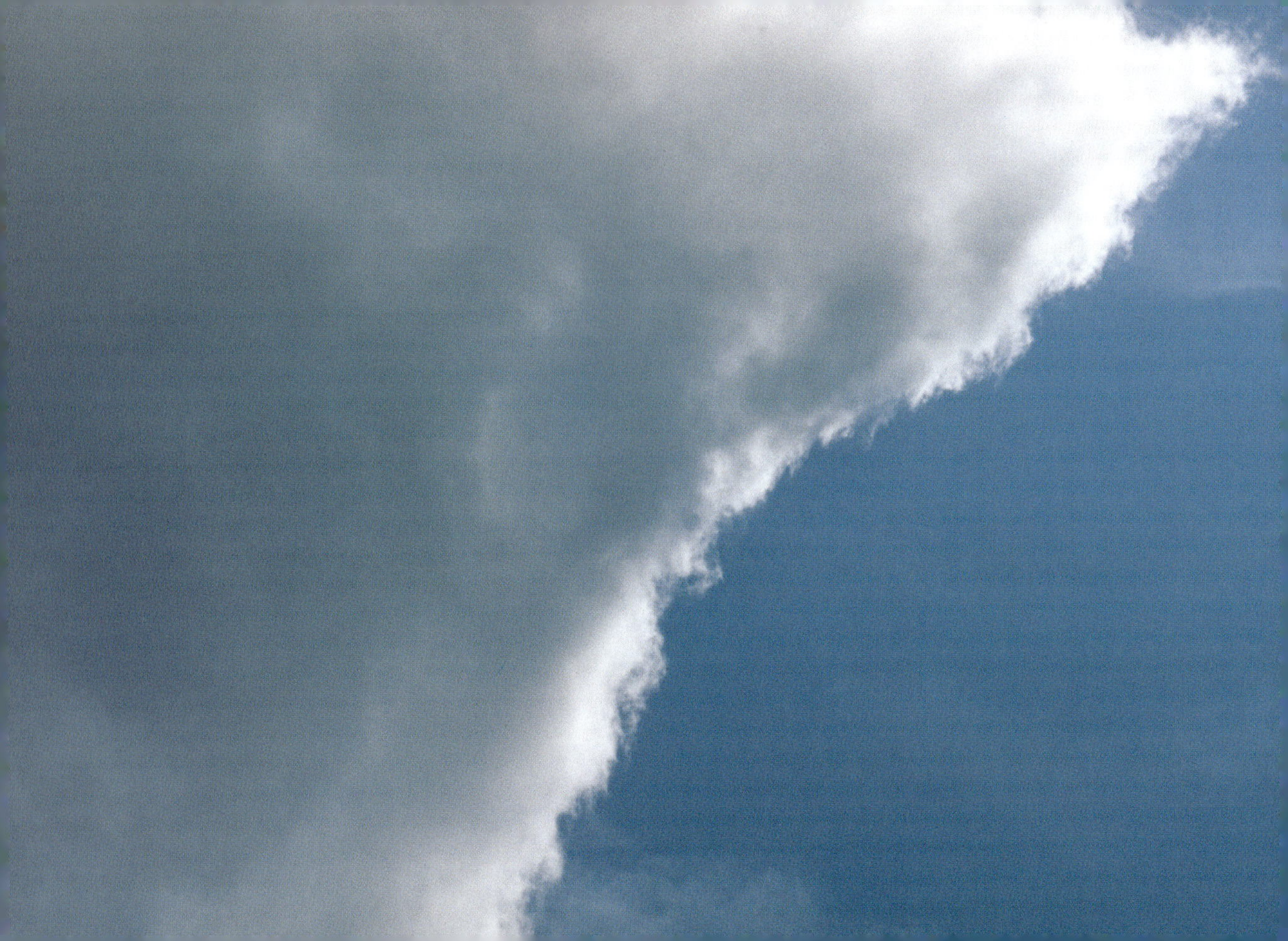

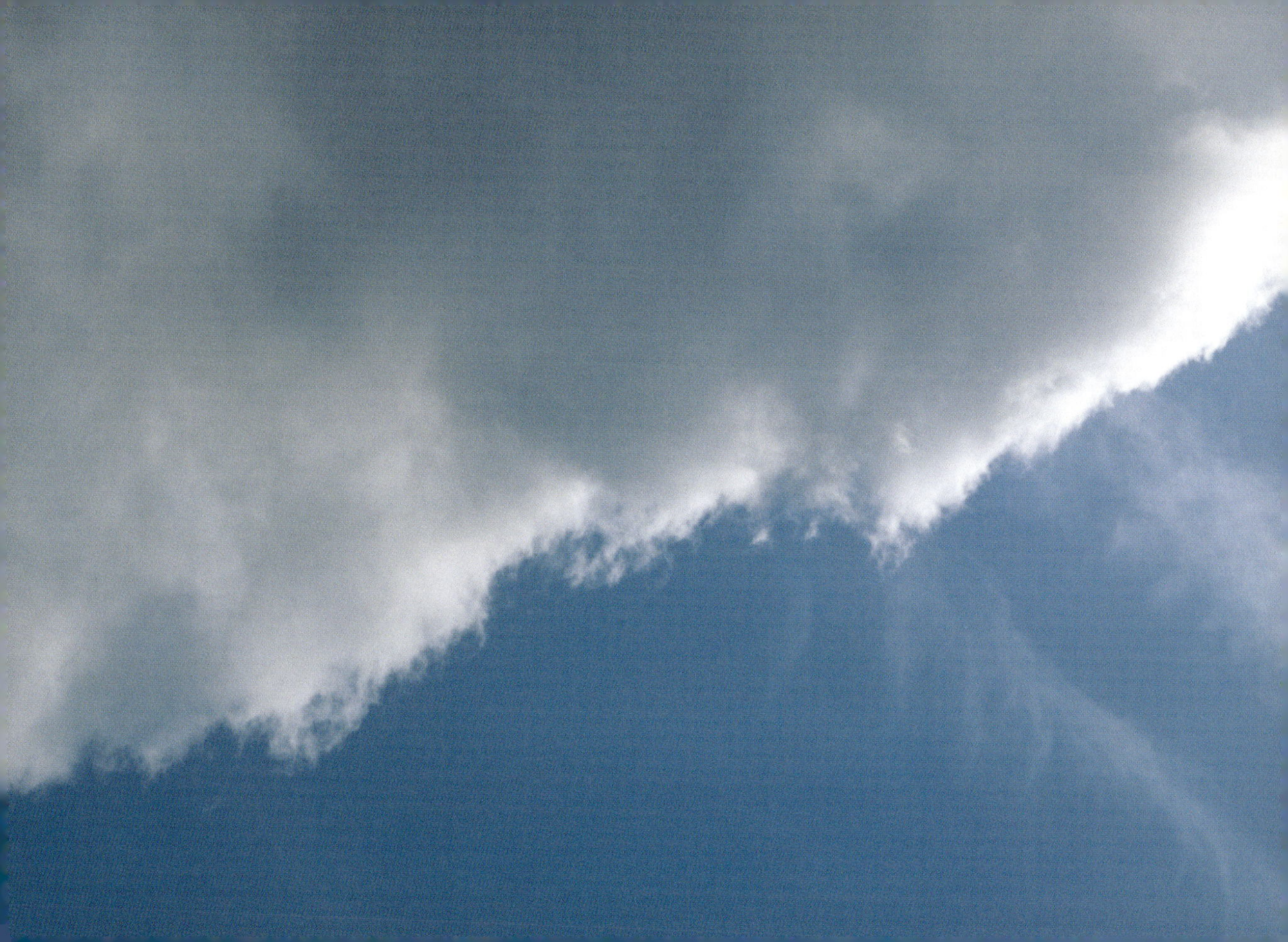

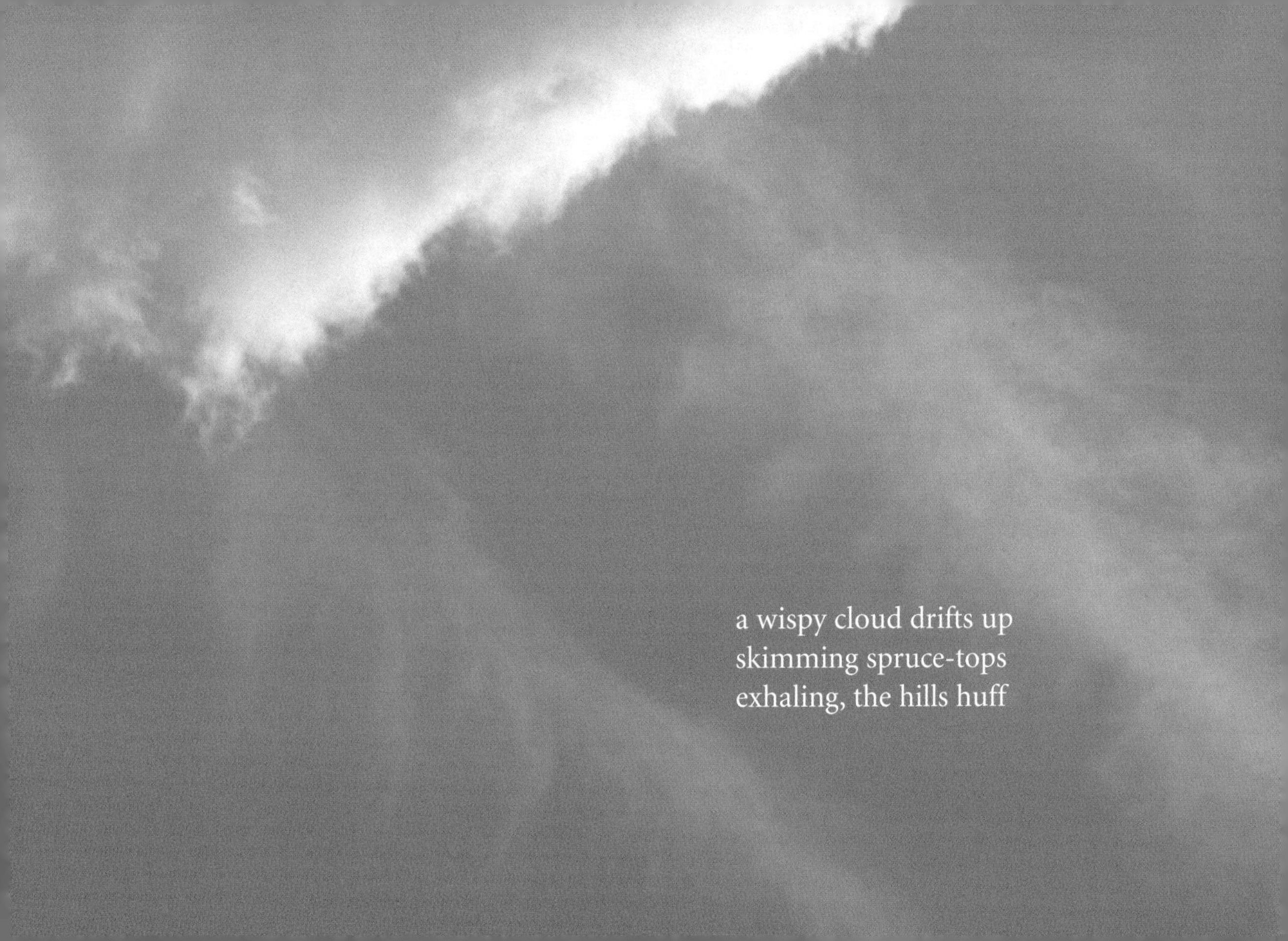

a wispy cloud drifts up
skimming spruce-tops
exhaling, the hills huff

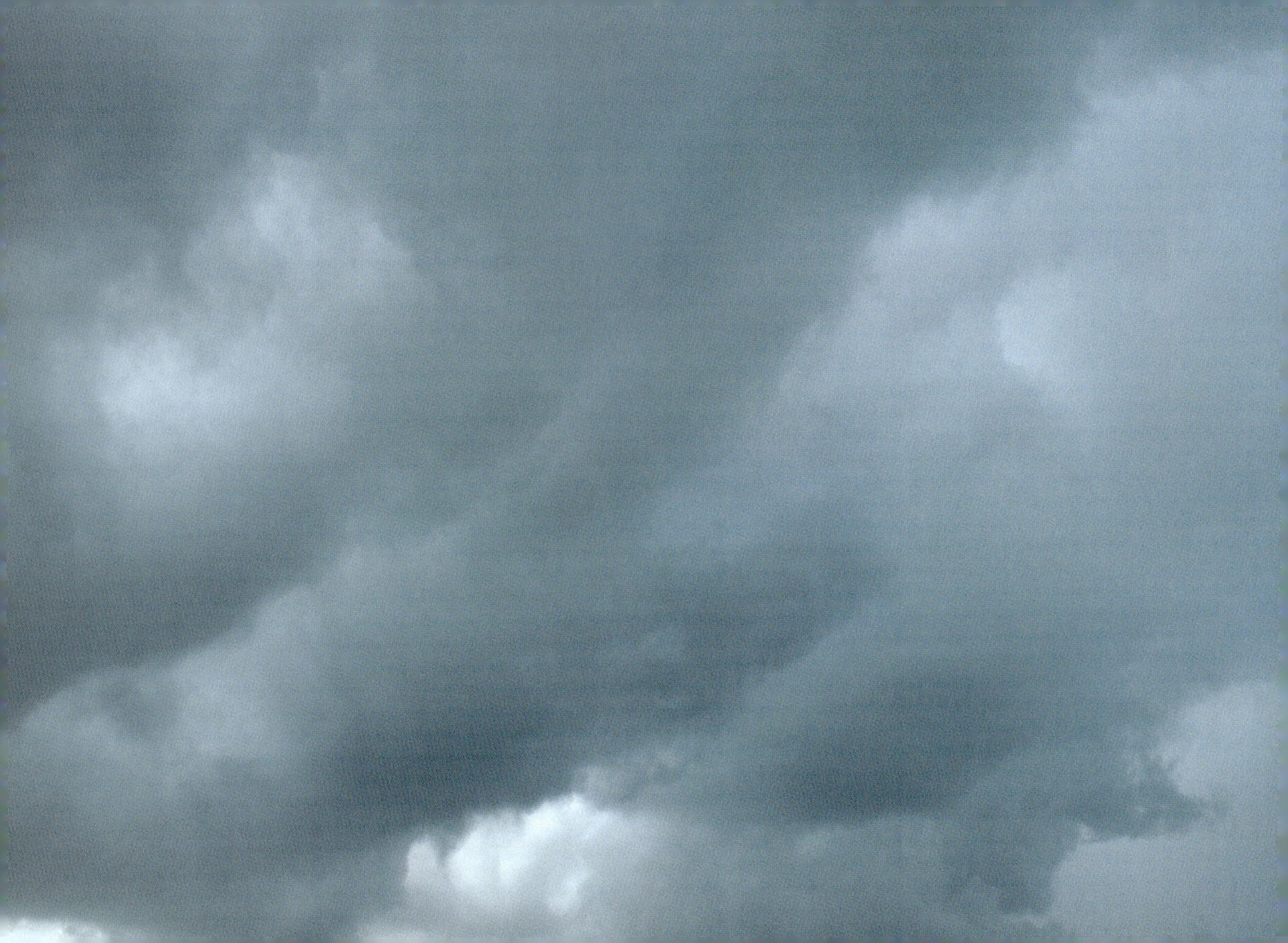

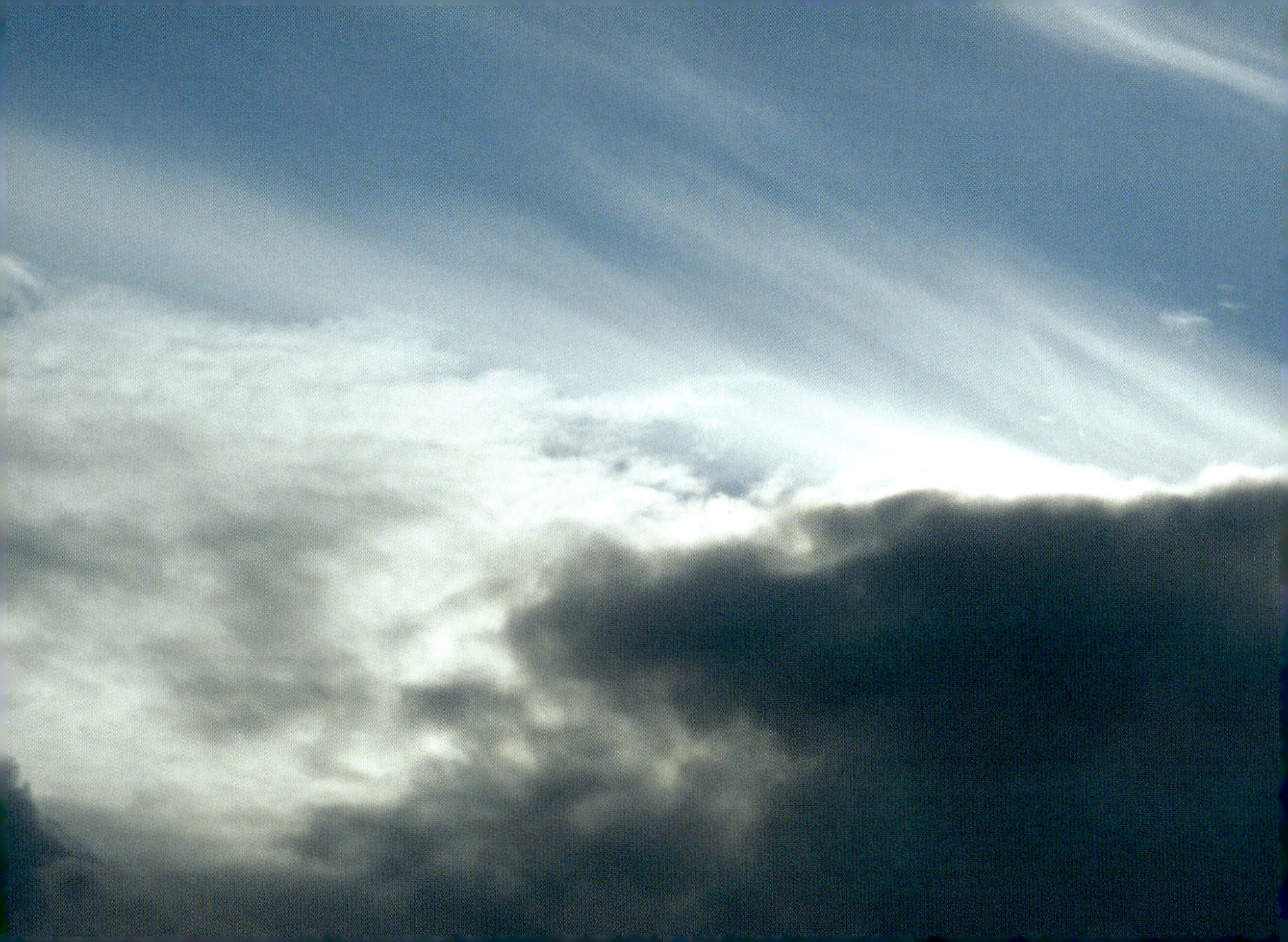

peak on peak
clouds gathered low
at the sea's horizon

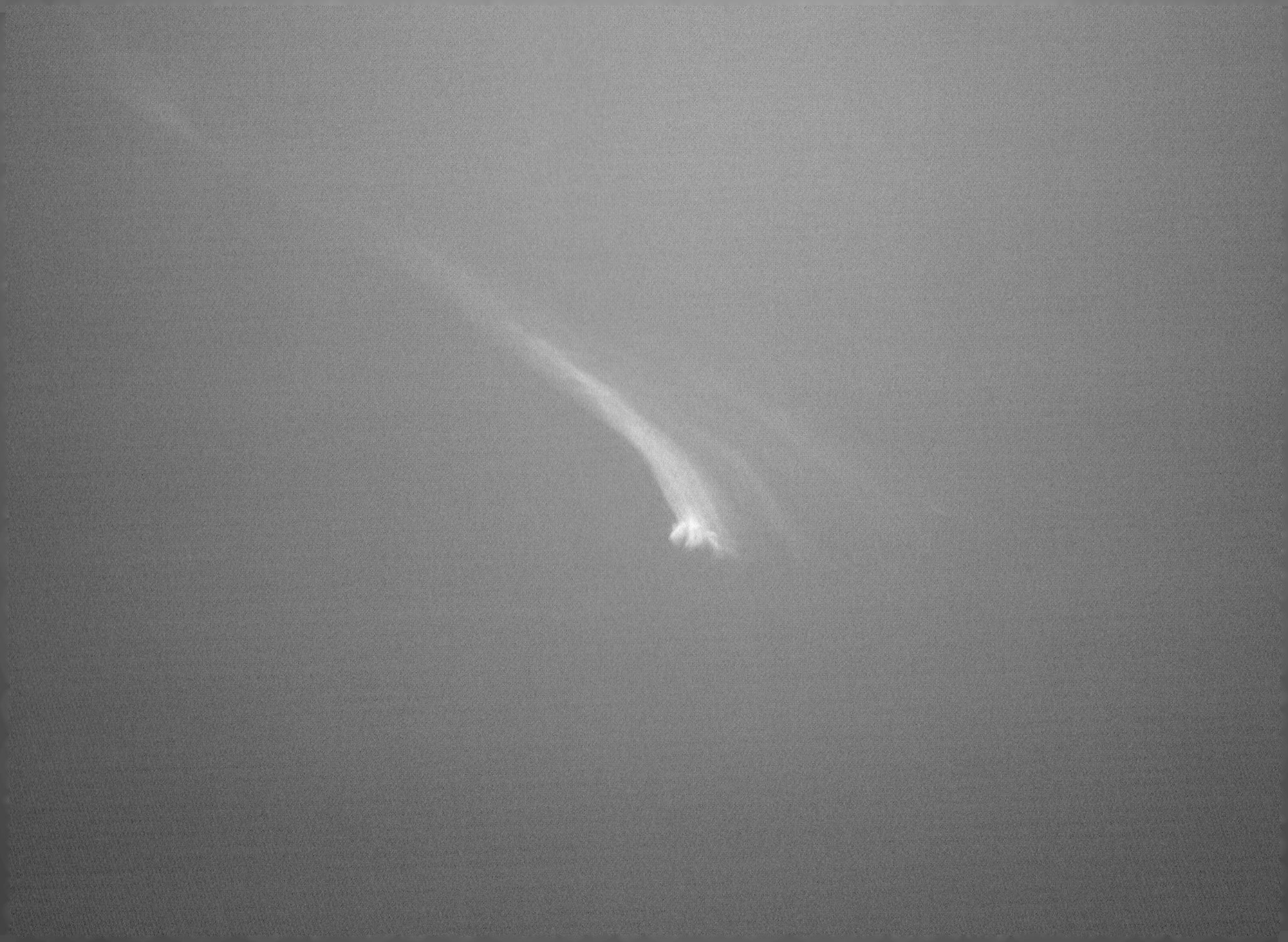

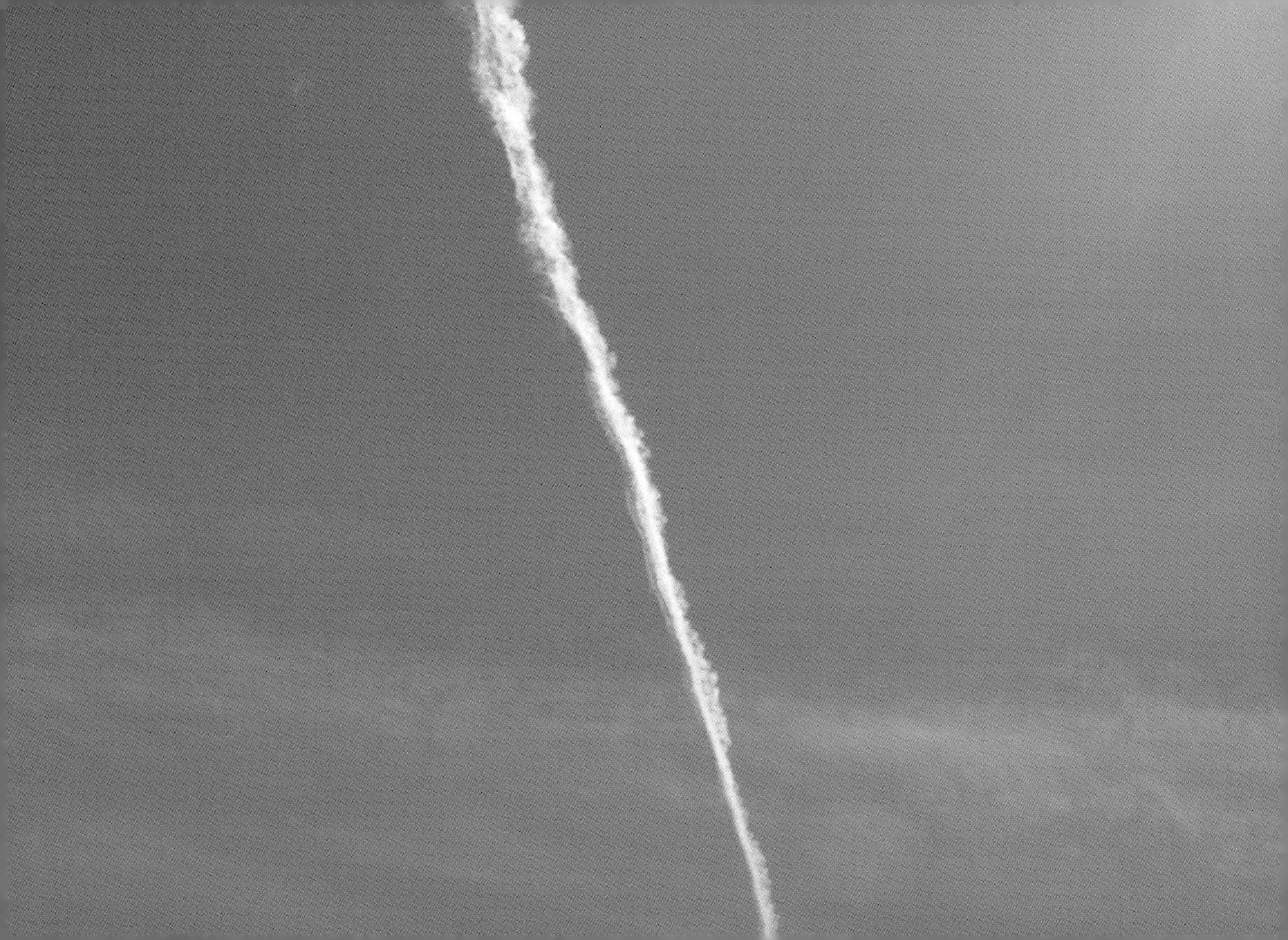

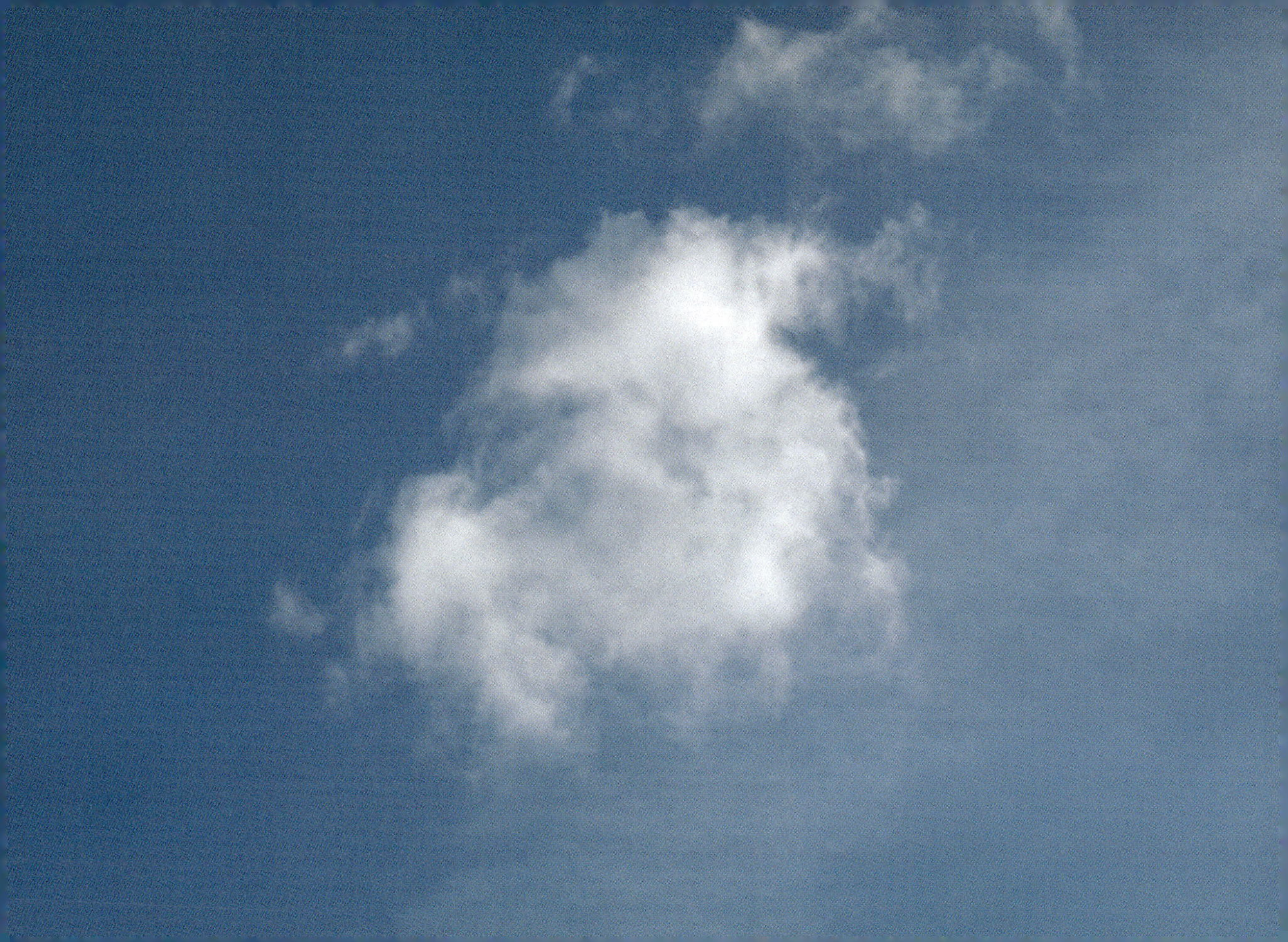

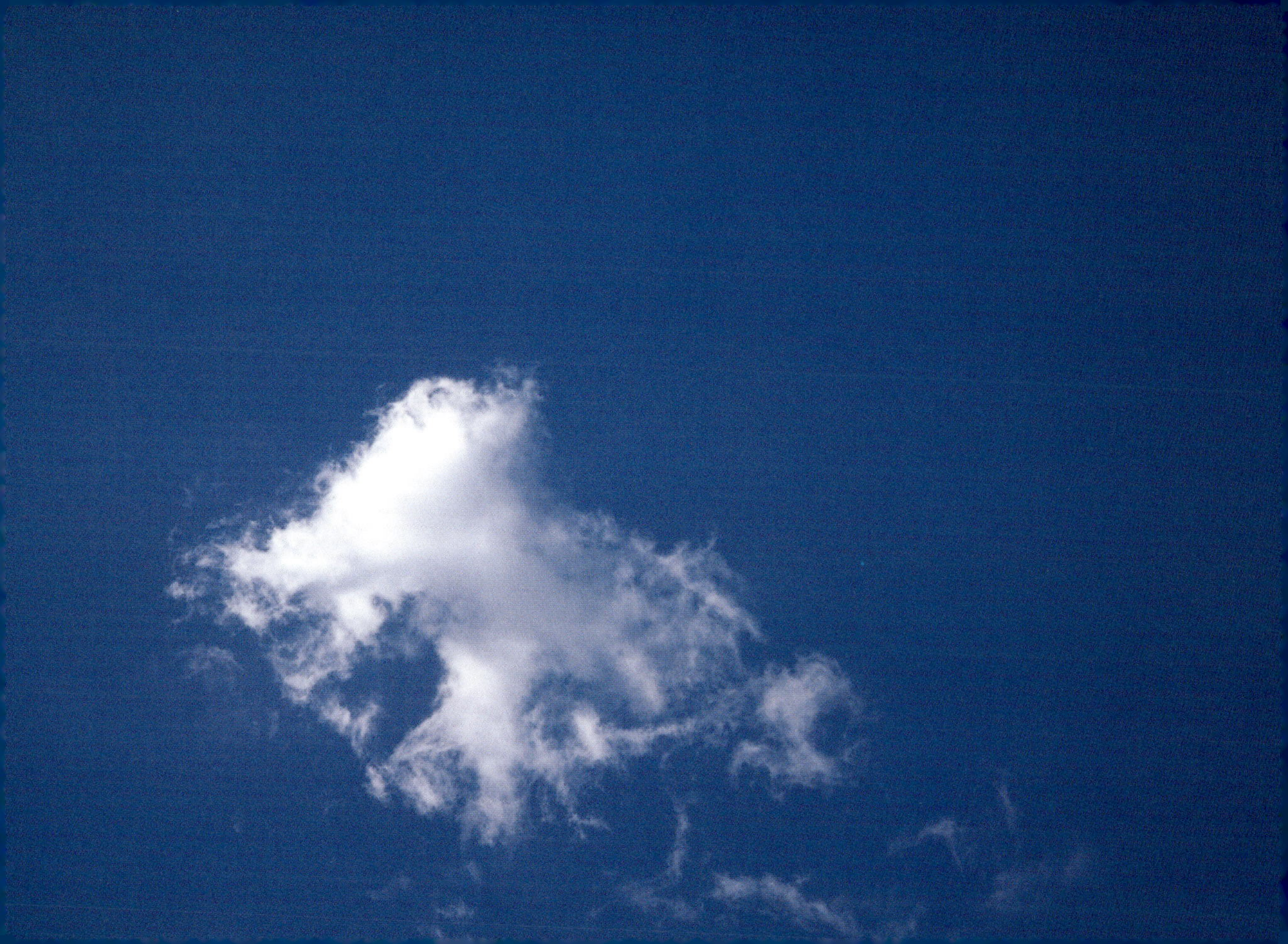

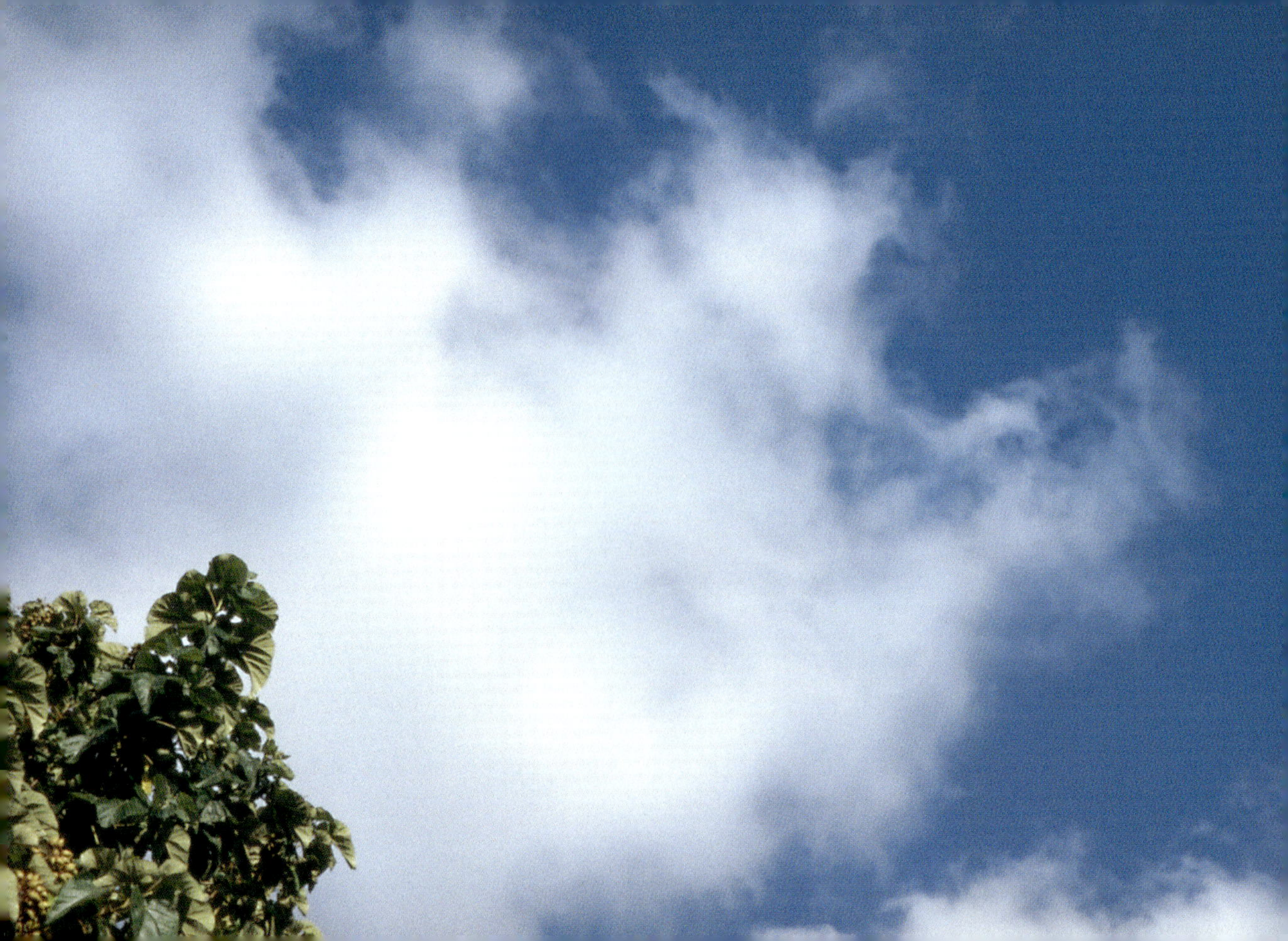

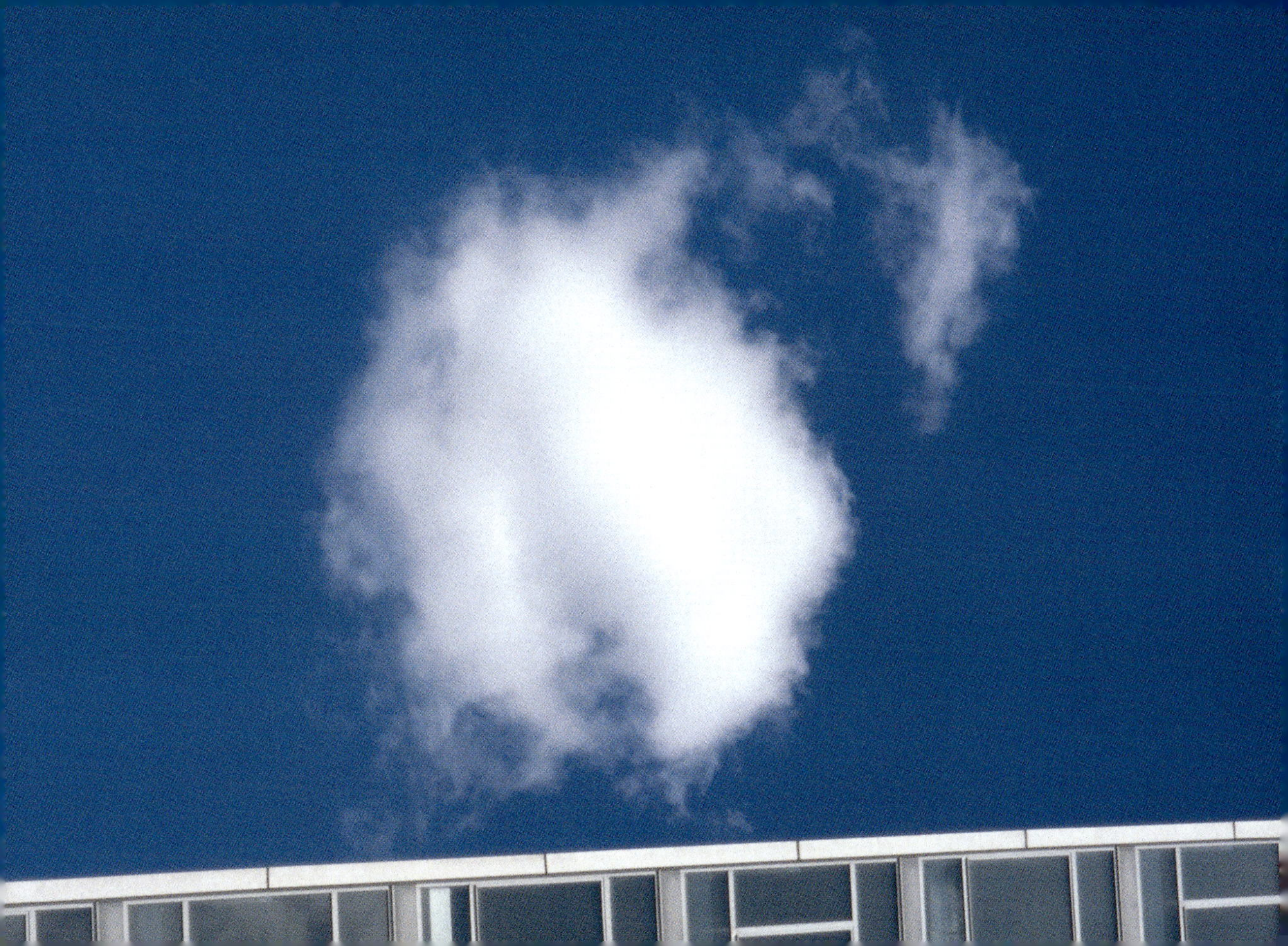

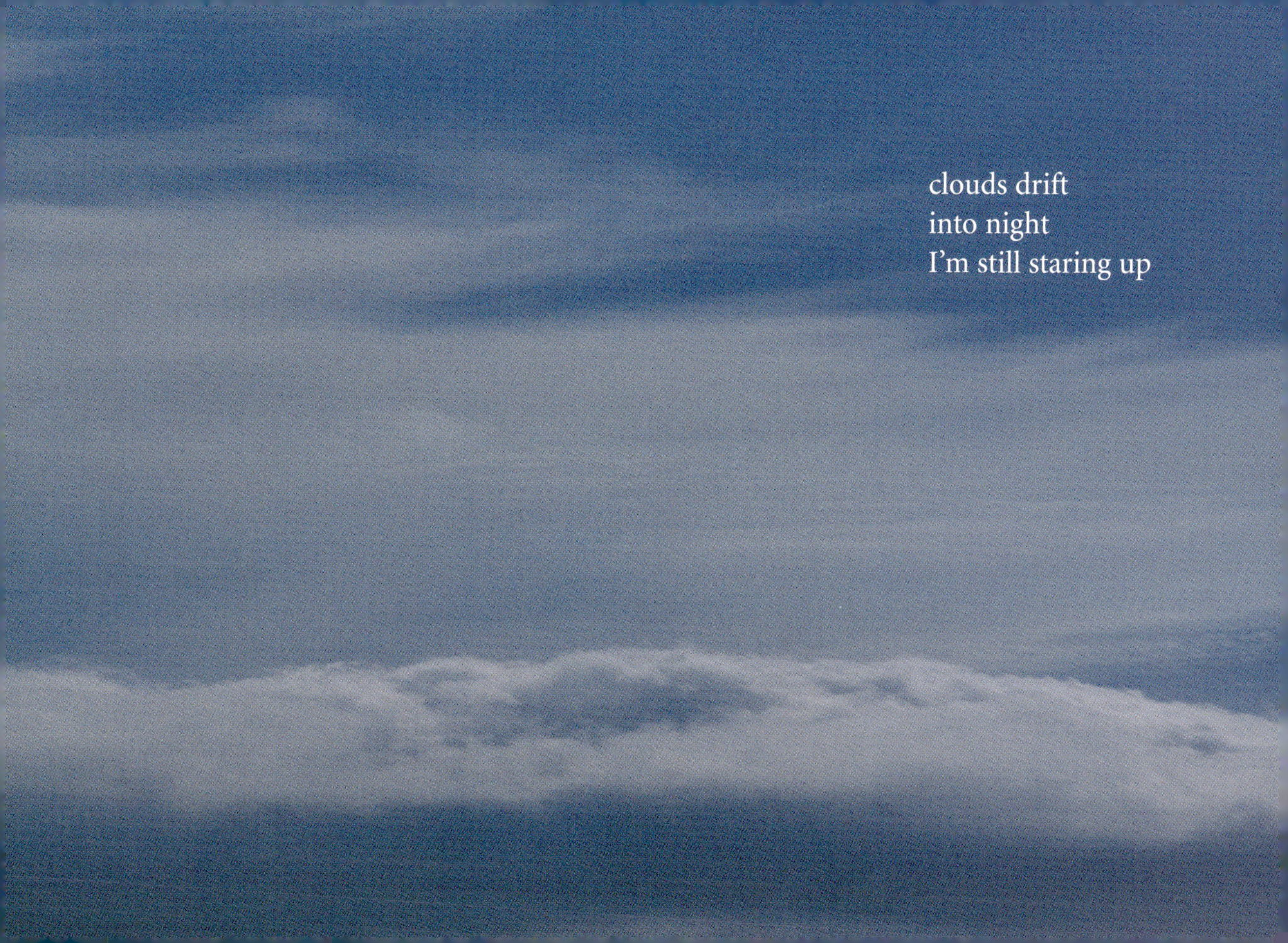

clouds drift
into night
I'm still staring up

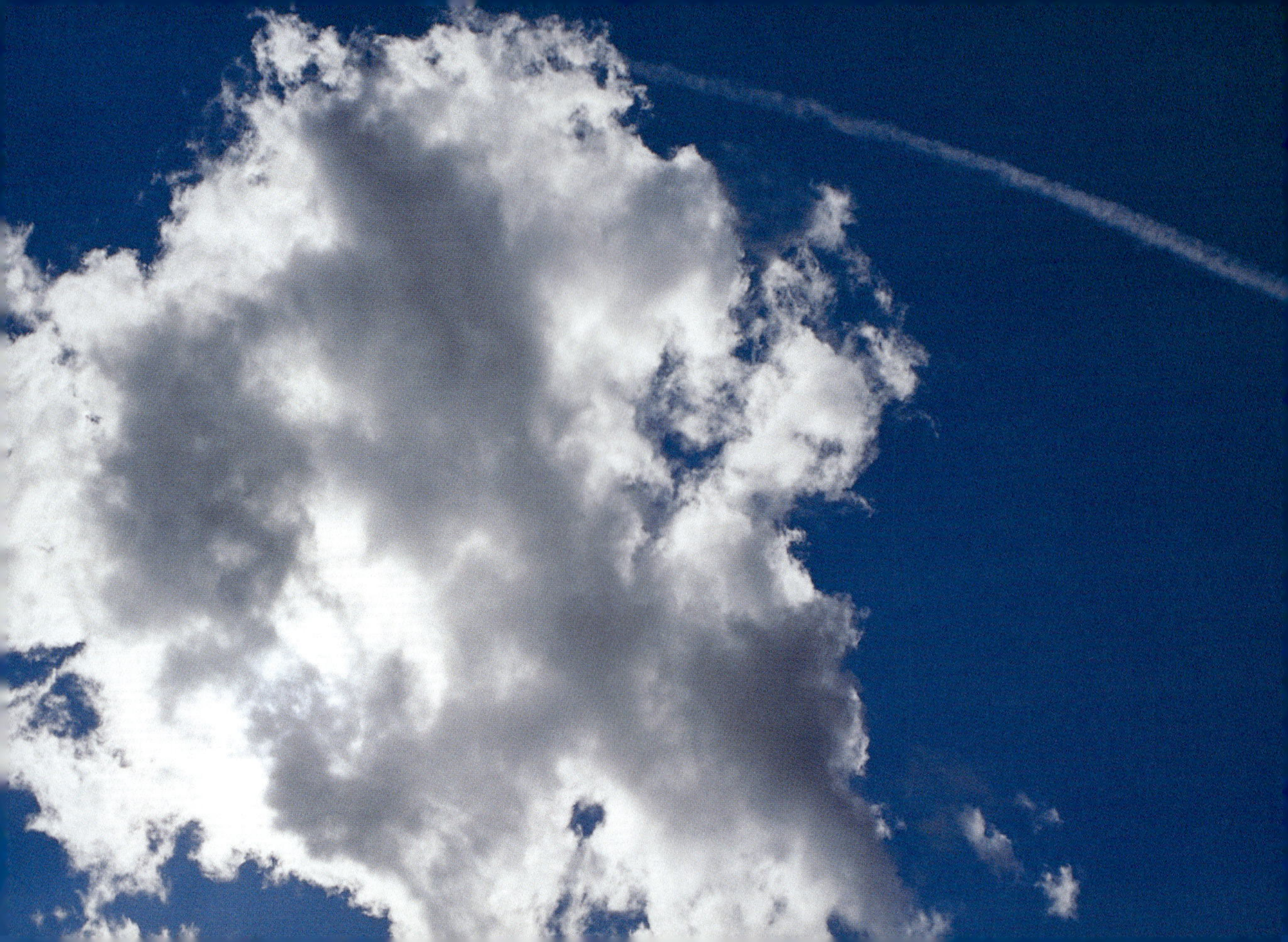

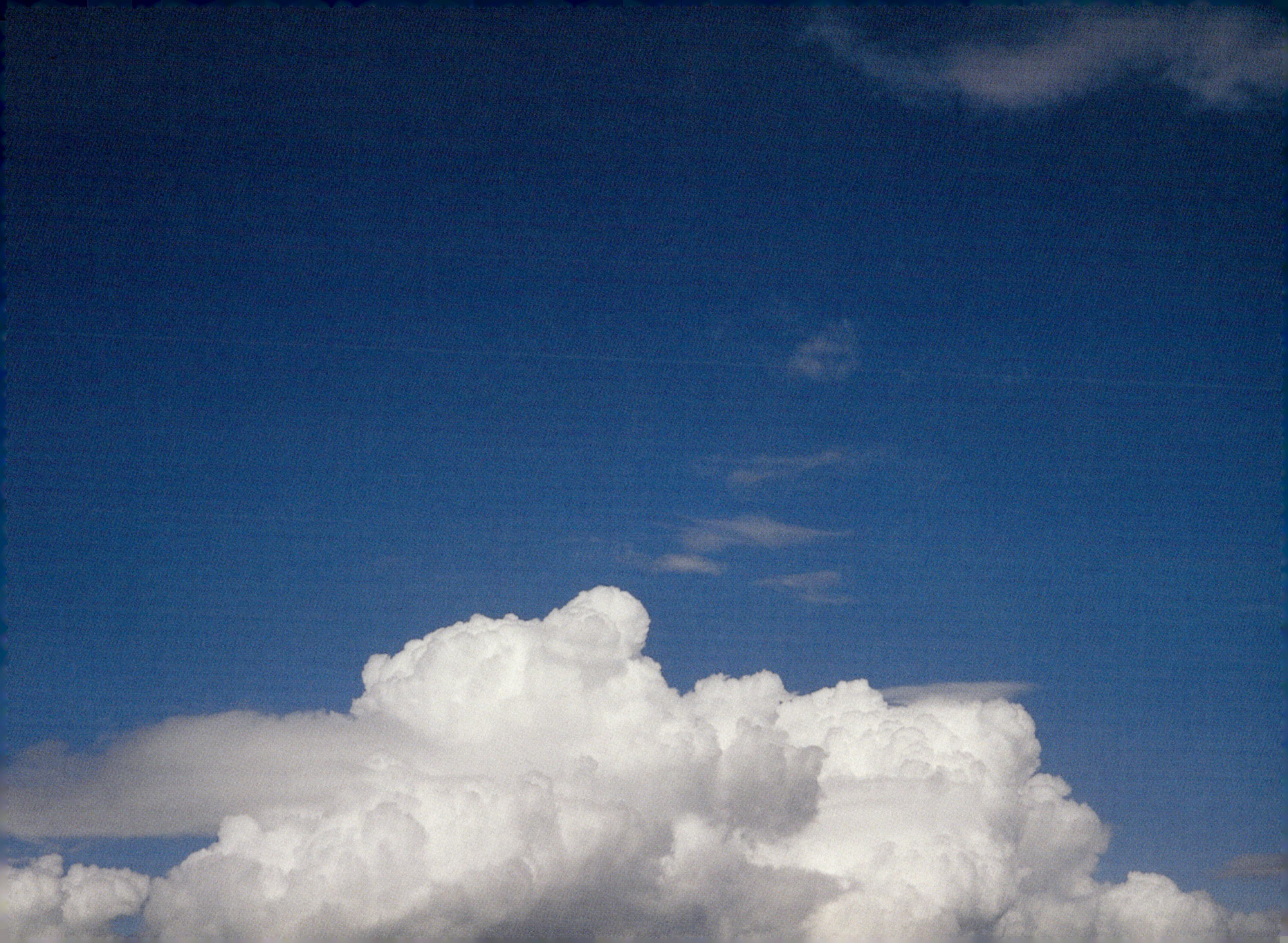

You are invited to contribute to an anthology of wind blown clouds.

Take a color slide photograph of a wind blown cloud. Include a note of
the date and place the slide was taken, your name, and mailing address.

There is no limit to the number of clouds that you can submit.
The clouds will be collected in an archive.

Alec Finlay

The Archive, Off Quay Building, Foundry Lane, Byker, Newcastle-upon-Tyne NE6 1LH

The Archive of
Wind Blown Clouds
List of Contributors

Director: Alec Finlay
Archivist: Laura Harrington

Catrin Anderson
Shelagh Atkinson
Robert Austin
Daniel Bell
Tim Bird
Angela Blacklock-Brown
Sarah Bloomfield
Sarah Bodman
Susan Brind
Alexander Braun
Celeste M. Brignac
Philip Clow
Lilian Cooper
Gerry Cordon
Lisa Curtiss
John Chapman
Ian Chesters
Steve Chettil
Noel Conor
John Cuthbert
Marie-Pierre d'Arifat
Kok de Haan
Caroline Dear

Kris Douglas
Helen Douglas
Jorn Ebner
Jan Eerala
Gerald England
Amparo Montero Espina
Bob Evans
Peter J. Evans
Michael Fairless
David Faithfull
Alec Finlay
Peter Foolen
Nayua Frangooli
Franziska Furter
Stephen Henry Gill
Malcolm Gibson
Maeve Gillies
Valerie Gillies
Raymond Godley
Morven Gregor
Emma Halliday
Jim Harold
Laura Harrington

Andrew Hodson

Vincent Hopson

Jamie House

Alexandra Hutchinson

Neal Johnstone

Mette Karlsvik

Yoon Jung Kim

Uta Kogelsberg

Evangelia Kokkinou

Yohan Lippens

Petra Liebentanz

Ann Lockey

H. B. Mackie

Catriona Macinnes

Fiona Macdonald Lockhart

Brent Mackey

Sandie Macrae

Heather MacNiven

Emil Jocobi Madson

Christopher Manson

Maris

Andrew Matthews

Guy Moreton

Rory Munro

Elspeth Murray

Brendon O'Brien

Mike Pearce

Rita Pearce

Alistair Peebles

Raymond N. Rayner

Karen Riveron

Olga Robertson

Gail Ronan

Brigid Sundaram

Telfer Stokes

Janet Scurr

Ruth Sheldon

Elane Small

Rose E. Steber

Phil Towle

Donald Urquhart

Natasha Werdmuller Von Elgg

David Williams

Ferelyth Wills

David Wood

Michael Wurstbuster

Index of poems

Acknowledgments

I would like to thank all of the contributors to the Archive, in particular Lillian Cooper, Amparo Montero Espina, Franziska Furter, Morven Gregor, Alexander and Susan Maris, and David Williams.

For their help in translating and collecting the haiku that accompany the photographs, my thanks go to Paul Conneally, Gerry Loose, and Yushin Toda. The translation from Basho's Oku *No Hosomichi* is by the late Cid Corman.

The project has proceeded with the support of a number of contemporary art institutions that have copublished invitation cards. These include BALTIC: The Centre for Contemporary Art, Gateshead (Sune Nordgren, Emma Thomas, Jude Watt); Brighton Photography Biennial (Jeremy Millar); Yorkshire Sculpture Park, West Bretton (Alex Hodby, Claire Lilley); Street Level Gallery, Glasgow (Malcolm Dickinson); Orchard Agency, Derry (Brendan McMenamin); Blaffer Museum, Houston (Gavin Morrison); Royal Scottish Academy, Edinburgh (Friends of the RSA, Colin Greenslade); STOT website (Kelly Richardson, www.stot.org). The postcards were designed by Cluny Sheeler (platform projects).

I would also like to thank everyone at Rizzoli who contributed to the publication of the book (Charles Miers, Eva Prinz, Ilaria Fusina, Ellen Nidy, and Anet Sirna-Bruder).

The completion of the book would not have been possible without the continuing support of Laura Harrington, Beth Rowson, and Lucy Richards.

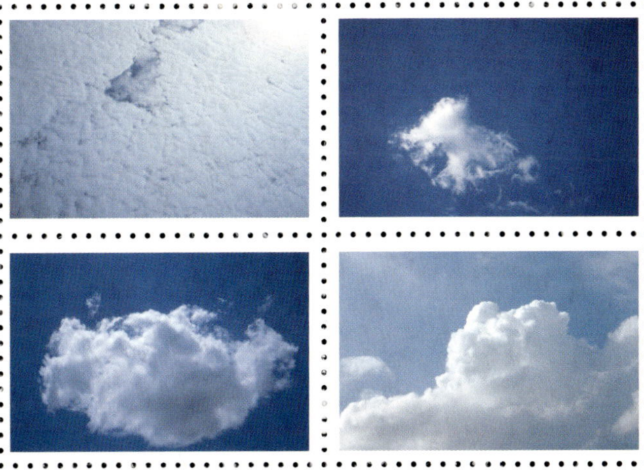